Nylon and Bombs

STUDIES IN INDUSTRY AND SOCIETY

Philip B. Scranton, Series Editor

 Published with the assistance of
the Hagley Museum and Library

Related titles in the series:

ELSPETH H. BROWN, *The Corporate Eye: Photography and the Rationalization of American Commercial Culture, 1884–1929*

CLARK DAVIS, *Company Men: White-Collar Life and Corporate Cultures in Los Angeles, 1892–1941*

PAMELA WALKER LAIRD, *Advertising Progress: American Business and the Rise of Consumer Marketing*

KAREN WARD MAHAR, *Women Filmmakers in Early Hollywood*

JOANNE YATES, *Control through Communication: The Rise of System in American Management*

JOANNE YATES, *Structuring the Information Age: Life Insurance and Technology in the Twentieth Century*

Nylon and Bombs

DuPont and the March of
Modern America

Pap A. Ndiaye

Translated by Elborg Forster

The Johns Hopkins University Press
Baltimore

Ouvrage publié avec le concours du Ministère français chargé de la culture—Centre National du Livre. (This book has been published with the assistance of the Center for the Study of the Book, a section of the French Ministry of Culture.)

The Johns Hopkins University Press
2715 North Charles Street
Baltimore, Maryland 21218-4363
www.press.jhu.edu

Library of Congress Cataloging-in-Publication Data

Ndiaye, Pap
     [Du nylon et des bombes. English]
     Nylon and bombs : Dupont and the march of modern America / Pap A.
Ndiaye ; translated by Elborg Forster.
          p.   cm. — (Studies in industry and society)
     Includes bibliographical references and index.
     ISBN 13: 978-0-8018-8444-3
     ISBN 10: 0-8018-8444-6 (hardcover : alk. paper)
     1. E.I. du Pont de Nemours & Company—History. 2. Chemical
industry—United States—History. 3. Research, Industrial—United States—
History. 4. Military—industrial complex—United States—History. I. Title.
II. Series
HD9651.9.D8N3513 2006
338.7'6600973—dc22   2006004143

A catalog record for this book is available from the British Library.

Contents

Illustrations appear following page 140.

Translator's Note

This book has—with the author's approval—undergone some changes in translation. Most important, a lengthy section of the original chapter 1, entitled "How to Write a History of the Chemical Industry," is incorporated into the "Essay on Sources and Historiography," which now takes the place of the "Bibliographical References" in the original text. The rest of chapter 1 became part of a new chapter 1, formerly chapter 2. Hence, the original chapter 3 is now chapter 2, and so forth.

In addition, American sources have sometimes been supplied in place of French sources, for instance, with respect to subjects like Prohibition and the Hanford environmental cleanup. Some references to French equivalents of American institutions have been omitted.

My translation has greatly benefited from the work of Peter Dreyer, the copy editor for the Johns Hopkins University Press. I thank him for many excellent suggestions; any remaining errors or infelicities of style are of course my own.

Elborg Forster

Nylon and Bombs

Introduction

In 1952, the editor and social historian Frederick Lewis Allen hailed the advent of a "responsible capitalism," which, jointly structured by government and big business, would be capable of meeting the challenges of a new era. Since the beginning of the century, he continued, big business had been changing. Companies were increasingly run by professionals, engineers, and managers who cared about the common good and worked hand in hand with government experts. While Allen worried about a materialistic American civilization—like Carthage, which "left no impact on the ages to follow it"—his words were nonetheless imbued with a characteristic optimism. He, and indeed most of his contemporaries, felt that the United States had entered a new era of political and social harmony sustained by abundance and protected by a nuclear shield. Allen pointed out that nylon stockings were produced at a rate of 543 million pairs per year, that is, "enough to provide every female in the country, from the age of 14 up, with between 9 and 10 pairs apiece"; "how is that for an example of the dynamic logic of mass production, producing luxury for all?"[1]

Today these words may appear terribly naive—a mark of self-satisfied innocence. But Allen clearly realized that he had to take the long view if he were to grasp the origins of American political and economic power at mid-century. He did not merely attribute it to the postwar boom and a Europe that was binding up its wounds. He also made clear that his version of modernization was predicated on the mass consumption of highly profitable goods. Allen was rather less voluble, however, when it came to another, darker but equally notable, facet of American life, namely, the rise of a permanent war economy centered on the mass production of atomic bombs and nurtured by the arms race between the Soviet Union and the United States.

This book explores the dual role one large American firm, E. I. Du Pont de Nemours and Company (DuPont),[2] played in both the steep rise in mass consumption (with its cultural ramifications) and the building of the notorious "military-industrial complex."

DuPont did indeed pioneer in both synthetic fibers (nylon) and nuclear components (plutonium). Silky, shiny, light, tough, crease-resistant, and infinitely versatile, nylon revolutionized the textile industry. It led to the creation of plastics, and became a part of our culture. In the 1950s, nylon symbolized a new way of life, the future, the spirit of America and its mythical modernity. "If it's nylon, it's prettier, and oh! how fast it dries!" boasted the advertisement. One of the century's most brilliant industrial achievements, nylon earned DuPont billions of dollars.

For its part, plutonium was one of the symbols of the half-century that was coming to an end, but is also of its darker side, the Cold War and the arms race. It was the product of a scientific and industrial collaboration between eminent physicists, military men, and the DuPont engineers who had earlier developed nylon. Quite aside from the case of DuPont, these aspects of twentieth-century American history have generally been dissociated and studied separately by historians, no doubt as a result of disciplinary logics and intellectual choices that did not favor analysis of civil and military activities in conjunction with each other. In contrast, this book studies them together. Moreover, one often feels intuitively that they were historically linked, although this remains to be demonstrated.

What, then, was the contribution of a great enterprise like DuPont to the genesis and the temporary dominance of the celebrated "American model" that allegedly yoked prosperity tightly together with security, especially in the years immediately following World War II? To address this question, one must begin by studying the history of a group of professionals: the chemical engineers of DuPont, their activities and their careers. These engineers are known to posterity for having developed nylon in the 1930s, and then the plutonium so crucial to the Manhattan Project. They changed the culture of their company, drawing it into the public sphere, while responding to a massive demand for everyday consumer goods. Their experience combined the history of professionalization with that of technology; it showed how chemical engineering came into being simultaneously in academia and industry and how a generation of young engineers made their presence increasingly felt. As engineers and managers, these men (there were practically no women among them) constructed a network that came to be known as the military-industrial complex, while supplying the civilian market with widely used high-technology products like nylon. When Allen wrote his essay, chemical engineers had become mass-production experts—apparently indispensable not only to the prosperity but to the security of the United States.

The task at hand, then, is to examine the interwar reorganization of DuPont that science and the dismantlement of workshop traditions prompted; the development of chemical engineering as a field; DuPont's hiring of a new generation of chemical engineers; and finally the invention of nylon, the epitome of this new culture. Then, beginning with World War II, we turn to DuPont's participation in the Manhattan Project and subsequent military nuclear projects, without losing sight of the fact that the company remained fundamentally committed to the civilian market. At the crossroads of different types of consumption, chemical engineers were in a strategic position to further their careers within a rejuvenated company.

Not long afterward, in the late 1960s and the 1970s, the chemical engineers' image became seriously tarnished when they were accused of having contributed to the devastation of the environment and the arms race. Even nylon fell from fashion. "People today find nylon tacky and crackly, dowdy and fun, just like the 1950s," a French woman journalist wrote in 1987.[3] But in the early 1950s, the majority of Americans still thought that the mass production of sophisticated goods under the apparently efficient and impartial aegis of engineers and managers had given them the best means of guaranteeing steady growth, defusing social tensions, and maintaining social peace.

The following pages examine DuPont in a relational perspective that includes its internal operations and functional responses to the market, but also how it interacted with other organizations, manipulated or attempted to manipulate the political and regulatory environment, and presented itself (and responded) to society at large. The great twentieth-century American corporation is viewed as a huge production machine, a knowledge magnet, a circulator of information, and a center of gravity in the network structured around the two poles of the market and the state.

This study proceeds only in part chronologically. Each chapter treats a specific theme. Chapter 1 deals with DuPont's early history, the evolution of chemical engineering, DuPont's "proto–chemical engineers," early experiments with poison gas for military use, and the establishment of training programs at the Massachusetts Institute of Technology and a few other universities between 1910 and 1920. Chapter 2 focuses on the emergence of chemical engineers within DuPont, from the early 1920s to the late 1930s, devoting special attention to the departments in which they had the greatest impact and to the early stages of high-pressure techniques that allowed them to strengthen their position within the corporation and eventually develop nylon. Chapter 3 looks at DuPont's unique business culture in the interwar years, when company directors engaged in a last-ditch fight against the New Deal, while also seeking to anticipate the needs of a growing consumer society. Chapter 4 examines DuPont's key role in the Manhattan Project, the secret program to build an atomic bomb, in which chemical engineers invested their organizational and technological know-how and joined forces with the federal government. Chapter 5 moves on to the postwar production and marketing of nylon, which multiplied DuPont's fortune and delighted its customers, and to the firm's simultaneous contribution to the nuclear arms buildup. By then, DuPont's chemical engineers had reached the height of their careers and come to represent the ideal of the politically neutral expert who stood calmly at the crossroads of high-volume military and civilian consumption, but starting in the mid 1960s, they began to resemble sorcerer's apprentices, rather than exemplars of modernity. Finally, the book's Conclusion reflects on how, ironically, the very transformation of American society by technology in the twentieth century itself eroded the confident utopian ideology of the modern industrial age.

CHAPTER ONE

DuPont and the Rise of Chemical Engineering

Unlike the electrical industry, the American chemical industry did not appear out of nowhere at the end of the nineteenth century. In the 1840s, a small chemical industry began to develop in Pennsylvania and New Jersey, in the prosperous region that stretches from New York to Philadelphia. Small factories provided intermediary chemicals such as sulfuric acid, soda, bromine, and chlorine for the paper, leather, textile, glass, and soap industries, which made rapid strides. A trend toward concentration that had begun in the 1870s resulted in the formation of a few large companies, most of which specialized in the manufacture of inorganic chemicals. Kalbfleisch and the Grasselli Chemical Company, which produced sulfuric acid (a basic staple of the chemical industry in the eighteenth and nineteenth centuries), merged with Nichols in 1899 and became the General Chemical Company. As for the other major category of chemicals, the alkalines (soda, potash, and nitrated fertilizers), they were produced notably by the Solvay Process Company, the Michigan Alkali Company, the Mathieson Alkali Company, and other firms, such as Dow Chemical, which manufactured chlorine and bromine out of brine.

All these producers of heavy chemicals were essentially working for the industrial market, which was then experiencing spectacular growth, and they often found financial backing from other industries, such as paper- and glass-making. Martha Trescott has stressed, perhaps with some exaggeration, that the American chemical industry was already a leader in the field before World War I.[1] In 1879, this industrial sector employed some 11,000 people, and its products were valued at $44 million. By 1914, these figures were respectively 45,000 employees and $221.5 million. In volume, the American chemical industry was ahead of its German competition, but it did not equal the latter in terms of value added or technological investment.

When the young Charles Reese, future director of chemical research at DuPont, returned from the prestigious Heidelberg University in 1886, his doctor's diploma in hand, he did not immediately find work commensurate with his capability: "Do you know, there was not any opening for a chemist, to speak of, in those days. It was very hard to get a job. There were very few chemical engineers, and outside of the sulphuric acid industry, the only industry that employed chemists was the iron and steel industry where [the chemist] simply stood in the laboratory and made tests."[2]

While it is true that by the time of World War I, the American chemical industry was in the lead in terms of volume, its production consisted essentially of inorganic chemicals with relatively low added value, unlike the German industry, which was very advanced in organic chemistry (dyes, pharmaceuticals). The Americans did not produce any synthetic ammonia, cracking [the process of refining heavy hydrocarbons] was still at the experimental stage, and petrochemistry was in its infancy. To be sure, some small firms were making light chemicals and the manufacture of pharmaceuticals had sprung up in the last years of the century (Upjohn and Abbott Labs in the 1880s), while Dow, Montsanto, and Merck were developing organic chemicals. But at the time, the key to organic chemistry was the field of synthetic dyes, where the Germans had a quasi-monopoly. Before World War I, the few American firms that manufactured dyes used intermediary materials imported from Germany.[3] As a result, the first years of the war were extremely difficult for the American textile industry, which had to make do with a very limited range of colors because of the blockade of Germany. Around 1916, it was even feared that American fashion would have to feature white only.

On 9 July 1916, the *Deutschland*, a specially outfitted German submarine carrying a full cargo of synthetic dyes, entered Baltimore harbor, having evaded the British blockade that interrupted commercial exchanges between Germany and third countries, among them the United States, which at this point was officially still neutral. The American press made a great deal of this exploit, congratulated the brave crew, and speculated about the possibilities of supplying the American market with dyes by submarine. This lasted until the United States entered the war and the rearmed *Deutschland* sank several American ships.[4]

By contrast, the United States excelled in the manufacture of heavy chemicals (sulfuric acid, potash, soda), and American chemical plants were conceived for this purpose. Consisting of large units designed for high-volume output, they favored continuous-flow production techniques.

This situation of the American chemical industry differed from the other new industry, the electrical industry, in two respects. First, the electrical industry was more concentrated. At the turn of the century, General Electric, Westinghouse, and American Telephone and Telegraph dominated this sector, notably through their control of patents. By absorbing the competition, buying up patents registered by independent inventors, systematic innovation, and rigorous management, the pre–World War I electrical industry had achieved a decided technological and capitalist advance over industrial chemistry.[5] From the very beginning, its technological development, more complex and costly than it was in the chemical industry, furthered a vast trend toward concentration, encouraged by the banks, making the electrical industry a textbook case of managerial innovation as analyzed by the business historian Alfred D. Chandler Jr.[6] By the eve of World War I, the American electrical industry towered over its European competitors.

Secondly, the chemical industry combined traditional artisanal practices inherited from the nineteenth century with new scientific knowledge born at the end of the century, for the most part in Europe, particularly in Germany. Industrial chemistry long preserved an artisanal aspect in which the chemist's flair, the "rule of thumb," counted for a great deal—as it does to this day. A good chemist is first of all an experimenter who knows how to "handle" things and in whom manual dexterity is not the least important qualification. In their autobiographical writings, the great chemists have always stressed their taste for the concrete, for experimentation, even tinkering, no doubt in order to set themselves apart from a kindred and more prestigious science, physics, which is more inclined to engage in theoretical work, and even to philosophical parables.

One chemical firm would nonetheless turn the American chemical industry upside down in the first decades of the twentieth century—DuPont, an old lady who had been around for a hundred years and found a second wind at the dawn of the new century.

DuPont had dominated the gunpowder and explosives sector for a century. In 1802, in the early years of the young American republic, at the instigation of Thomas Jefferson, a gunpowder factory had been established on the banks of the Brandywine River near Wilmington, Delaware, by Eleuthère Irénée du Pont (1771–1834), a former student of the famous chemist Antoine Lavoisier's.[7] Starting in 1804, this concern, known as E. I. du Pont de Nemours and Company, supplied black powder (a mixture of saltpeter, charcoal, and sulfur) not only to the U.S. military but also to frontiersmen for hunting and clearing the land, and to mine owners and public works projects for blasting rock.[8] DuPont is one of the oldest American companies still in existence, along with a few banks and insurance companies, such as Bank of New York and Cigna, and probably the oldest industrial firm. Among DuPont's first employees were stonemasons, who had built its workshops and storage facilities out of solid granite. DuPont continued this practice of taking charge of the construction of new production units into the twentieth century. Despite some ups and downs, the company flourished, particularly under the leadership of its founder's grandson Lammot du Pont (1831–84), and it supplied the Union Army with high-quality gunpowder during the Civil War, when two regiments protected Wilmington from any Confederate incursion. Even then, the press did not fail to point out that DuPont made too much money from its gunpowder, for which it charged the Army a high price—an accusation that recurred again and again in the firm's history. In the 1880s, DuPont launched its dynamite manufacture at its Repauno plant in New Jersey, and also began to produce smokeless gunpowder for military use. Dynamite, which had been invented in Europe in 1866 by the Swedish chemist Alfred Nobel, was in great demand at the time, for it did wonders when it came to excavating for the foundations of the first skyscrapers and the New York subway, to cement making, and to large-scale coal mining.

The production of these new nitrate-based explosives soon made it necessary to hire chemists and technicians who could make improvements in manufacturing processes, as well as workers' safety. From the very beginning—and this was without question one of the structural reasons for its success—DuPont devoted particular care to safety and produced high-quality gunpowder. When the firm went into the manufacture of more dangerous commodities such as nitroglycerine and dynamite, it already had a solid tradition of stressing safety, although this did not prevent all accidents. In 1884, Lammot du Pont was killed in the explosion of the nitrate storage facility. After Lammot's two successors, Henry and Eugene, died prematurely, in 1889 and 1902, the direct heirs decided to sell the business to the competing Laflin & Rand Powder Company.

An important turning point was soon reached, when three young du Pont cousins, T. Coleman, Pierre S., and Alfred I. du Pont, bought back the family enterprise in 1902 and undertook to restart and restructure it.[9] At that time, DuPont was associated with the other manufacturers of gunpowder and explosives in two cartels, that is, two horizontal combines.[10] The new directors replaced the cartels with one integrated corporation, E. I du Pont de Nemours and Company, divided into separate departments and controlled by an Executive Committee (see fig. 1.1).

At this point, the firm still manufactured only gunpowder and explosives. But before long DuPont's new directors engaged the company in a whole range of products that could make use of cellulose and nitric acid, taking over other firms in order to do so. The reasons for this diversification are fairly well known. First of all, especially at a time when the U.S. Army and the Navy were setting up their own powder mills, the company no longer wanted to be dependent on military orders, which had hitherto essentially been its bread and butter. Moreover, the Sherman anti-trust legislation represented a serious threat to DuPont's quasi-monopoly on explosives, a threat that became concrete in 1911, when a court ruling forced DuPont to give up certain of its production units, which subsequently became the Hercules and Atlas Powder Companies.

For these reasons, in 1904, DuPont acquired the International Smokeless Powder and Solvents Company, which produced lacquers made from nitrocellulose, and six years later the Fabrikoid Company, which manufactured a kind of synthetic leather out of cotton and nitrocellulose, and later also the Arlington Company, the Fairfield Rubber Company, and Harrison Brothers Paint Company, a major manufacturer of paints. This process of diversification has been placed into a theoretical framework by Alfred Chandler, who dubbed it an "economy of scope," by which he means diversification founded on one basic product (in this case, cellulose). By the eve of the Great War, DuPont was already a diversified chemical business—even if more than 90 percent of its turnover involved gunpowder and explosives. This diversification at the beginning of the twentieth century marked the first stage in DuPont's opening up after its initial cartellization.

Shortly before the Civil War, Lammot du Pont, a graduate in chemistry

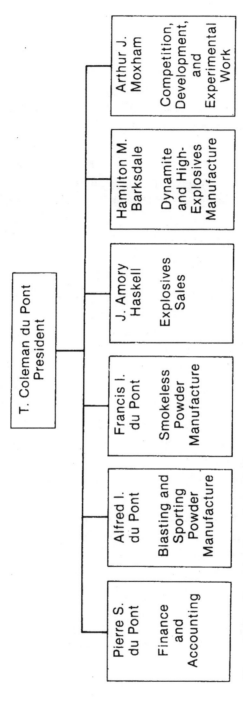

Fig. 1.1. DuPont Executive Committee roles in 1903

Source: David A. Hounshell and John Kenly Smith Jr., *Science and Corporate Strategy: Du Pont R&D, 1902–1980* (New York: Cambridge University Press, 1988), fig. 1.1 (p. 19).

of the University of Pennsylvania, had developed a gunpowder more efficient than the traditional black powder, but this did not mean that the company needed to employ high-powered scientists and technicians. By the early twentieth century, however, it was no longer possible to rely solely on the know-how of the firm's workforce. From now on, it needed experts trained in chemistry. The task was no longer, as it had been in the past, simply to produce gunpowder in large quantities, a process that did not require much technological and scientific knowledge, even though it demanded a firm grasp of the safe manufacture, handling, and storing of explosive materials.

The invention of dynamite by Alfred Nobel in 1866 radically changed the explosives industry, in Europe as in the United States. When DuPont began to produce dynamite in 1880, it had to hire chemists to perfect the rather delicate manufacturing processes, and to limit polluting acid by-products, which were killing the fish in the Delaware River. Hence the creation in 1902 of the first DuPont laboratory, the Eastern Laboratory, which was given the mission of improving the company's explosives and their manufacturing processes. This was followed in 1903 by the Experimental Station, which had a more general mission, in response to the du Pont cousins and their associates' desire for diversified activities. The Experimental Station was asked to study possible uses of nitrocellulose and to improve the products and the manufacturing processes of the many companies that DuPont had acquired. The historians David Hounshell and John K. Smith have provided detailed descriptions of the successive reorganizations of the laboratories in keeping with the company's general strategy, and particularly its efforts at diversification. But we must not get ahead of the story: in these first years, DuPont employed scarcely more than a dozen chemists, who continued to work with modest resources.[11]

The second consequence of DuPont's diversification concerned its manufacturing processes and the production of more sophisticated chemicals than in the past. This called for engineers capable of scaling up the work of the chemists' laboratory to industrial production. The question of how a formal body of scientific knowledge was grafted onto old experimental practices, and why it was that the late nineteenth century brought a changeover from the search for a theoretical basis for chemistry to a more empirical and concrete science, has occupied historians of chemistry for some time.[12] The main point to make here is that working in chemistry, more than with electricity, called for combining scientific knowledge with experimental technique. This hybrid character of industrial chemistry was to favor negotiations and exchanges of men and technical know-how between institutions of production and those engaged in developing technological information.

Although when the du Pont cousins took over the chemical firm and began to develop a strategy of diversification—when the need for a skilled labor force was not yet clearly stated and DuPont was still largely a company of "powdermen"— the outlines of career possibilities for a generation of new men were already becoming visible in a field practitioners came to call chemical engineering.

The Birth of Chemical Engineering

But what is a chemical engineer? Until the end of the first decade of the twentieth century, the answer was not self-evident. This profession was not mechanically born out of the rapid development of the chemical industry in nineteenth-century Europe and America. Professional compartmentalization is the result of social judgments rather than of a technical logic, which itself is in any case never unrelated to social circumstances. In the United States, chemical engineering was the fruit of a specific history, of projects that brought together partners from industry and the universities according to modalities that existed only in that country. Chemical engineering became a specifically American discipline; there were chemical engineers elsewhere in the world, and particularly in Europe, to be sure, but for a long time, there was a specifically American kind of chemical engineering, both in its theoretical foundations and in its practical know-how. It arose out of technical, professional, and social arrangements that might have been different.[13]

In the nineteenth century, American engineers were usually trained through apprenticeship, as were their British colleagues. Only 30 percent of those born before 1830 received university training, a proportion comparable to that in medicine.[14] The sociologist Paul Starr has shown, too, that in the nineteenth century, being a physician did not confer a clear and unequivocal position in American society. The very great disparities among physicians were a fairly accurate reflection of the country's social structure.[15] As in the case of the physician, the status of the engineer was defined not so much by his work as by his social background.

Future mechanical engineers, for instance, most often started their careers by working as apprentices in a shop, then became machine operators, and finally engineers.[16] The cases of Frederick Taylor and William Sellers offer useful examples, for these sons of prominent Philadelphia Quaker families began as apprentices in machine-tool firms before they became famous mechanical engineers.[17] More than anything else, Taylor and Sellers were brilliant tinkerers, self-taught practitioners working with largely empirical knowledge, who did not know much about the mechanical physics that was the core of their French colleagues' training. In an essay of 1908, Taylor pointed out that the engineers with university degrees had neither the strength of character nor the competence of those who, like himself, had dirtied their hands on the shop floor.[18] And when in 1882, he obtained a diploma in mechanical engineering from the Stevens Institute of Technology in Hoboken, New Jersey, one of those technical institutes that were springing up everywhere at that time, it was thanks to special arrangements that exempted him from examinations in thermodynamics and chemistry, although he also demonstrated remarkable competence in the mechanics of forging presses and machine tools.

It is true that a more theoretical approach, inspired by the French model, was being used in certain institutions, particularly the military academies, West Point for civil engineering and the U.S. Naval Academy in Annapolis, Maryland, for mechanical engineering. French military engineers, some twenty of whom had served in the young American army during the Revolutionary War, had inspired Congress to create a Corps of Engineers and Artillerists at West Point in upstate New York in 1794,[19] leading to the founding of the West Point Military Academy in 1802. Subsequently, beginning in the 1820s, the building of roads and canals prompted the states to hire civil engineers, most of whom were former military engineers. They were men like Claudius Crozet, a French artillery officer of Napoleon's Grande Armée, who emigrated to the United States, became an instructor at West Point, and then chief engineer for the state of Virginia in 1823.[20] Crozet redesigned the network of roads in Virginia and at the same time established rigorous technical norms for the construction of roads, bridges, and tunnels. And while the U.S. Naval Academy, founded in 1845, did not quite play the role for mechanical engineering that West Point had played for civil engineering, by the 1860s, the naval officers did begin to teach courses in the mechanics of the steam engine and other advanced subjects that were not offered at any other American university at the time.[21]

Many of these officers in mechanical engineering, like their West Point counterparts in civil engineering, were recruited by the numerous universities that were founded in the 1860s, thanks to the Morrill Act of 1862, which granted federal land for the creation of public institutions of higher learning to the states.[22] Moreover, by the middle of the century, the opening of engineering departments at major universities and the creation of engineering schools such as the Massachusetts Institute of Technology (1861) contributed to the scientific foundations of the training of engineers. In 1860, Yale University inaugurated its department of science and technology, the Sheffield Scientific School, which awarded the first American doctorate in chemistry.[23]

But for all that, the British influence—the tradition of the engineer trained by apprenticeship—remained strong, even if American engineers already had their own style. Several historians have pointed out that the Americans insisted on economy in construction and on lowering labor costs, in contrast to European engineers, for whom the elegance and the solidity of their accomplishments were more important.[24]

American engineers approached problems differently from graduates of the leading institutions of scientific education in France, whose theoretical and mathematical training, which had taken place under the aegis of a strong and centralized state, was more advanced than elsewhere. But in the late nineteenth century, some French municipalities and entrepreneurs began to create specialized engineering schools that looked toward industrial production rather than public service, and whose graduates had begun modestly by wearing the blue overalls of the worker or the white smock of the draftsman.[25] One should not, therefore, overstate the contrast between French and American engineers.

At the end of the nineteenth century, however, the so-called second industrial revolution favored a break with the empirical tradition treasured by the British and American engineers, and the chemical and electrical industries were essential factors in the industrial upheavals that occurred at that time. The new chemical industry was based, albeit to a lesser degree than its electrical counterpart, on advanced scientific and technical, rather than artisanal, knowledge, which became a material reality through registered patents (which doubled in number between 1866 and 1896). The electrical industry, in particular, was in the vanguard of industrial innovation between 1880 and 1920. General Electric, Westinghouse, and AT&T were in a position to integrate research activities into their respective organizations rather than entrust them to individual outside inventors or consulting engineers. Their directors felt that the growth of their companies was contingent on modern technology and on those who could apply it. The development of more and more complex technical systems, such as electric and telephone networks or continuous-flow chemical processes, called for a growing number of engineers who could design them and make them work. These companies attempted to raise the threshold of entry into the market high enough to discourage the competition, with respect both to capital investments and to technology.[26]

Two kinds of engineering specialties emerged in the wake of these technologically advanced industries and took their places alongside civil, mining, and mechanical engineering. They were practiced by electrical and chemical engineers, whose training combined the traditional know-how of the mechanical engineer with new scientific knowledge based on recent advances in physics and chemistry. In some ways, these new specialists formed the avant-garde of the scientific and technical revolution that in the following decades was to spread to American industry as a whole. Between 1880 and 1920, the number of engineers jumped from 7,000 to 136,000, an increase of 2,000 percent.[27] It should be noted in passing that scientific methods were not always beneficial to the various engineering specialties. Early in the century, the Bureau of Public Roads, a federal agency in charge of building highways, abandoned its traditional empirical methods of evaluating roadbeds in favor of a new scientific approach, which in the end turned out to be less effective.[28] Moreover, training by apprenticeship persisted in industries such as automobile manufacturing until the middle of the twentieth century. At the Ford Motor Company, in particular, the mechanical wizards around the boss had never been to college—and were proud of it.[29] Robert McNamara, the former secretary of defense, recalls in his memoirs that when he went to work at Ford in 1946, together with some other young men sporting brand-new diplomas (the famous "whiz kids"), only a handful of the 1,000 high-level managers of the automobile firm were university graduates.[30]

In short, it is important not to confuse the history of technology with that of innovation.[31] Widespread technical practices and old professional distinctions were not necessarily swept away by "economic modernization," to use a well-known term that has obscured historical understanding, and that should be scrapped by today's historians of technology. But chemical and electrical engineering could not get along without a certain minimum theoretical knowledge in physical chemistry and mathematical physics.

The Proto–Chemical Engineers

The term *chemical engineer* probably appeared for the first time in 1839 in England in the *Dictionary of the Arts, Manufactures and Mines*, at a point when great strides were being made by the European chemical industry, particularly in Great Britain and France. At that time, it referred to technicians with a rather limited knowledge of chemistry, often mechanical engineers working in chemical factories. The term *engineer* then meant a resourceful mechanic with practical know-how, who performed an essential function in the production process, rather than someone who had earned a diploma by undergoing rigorous formal training. Until the end of the nineteenth century, the definition and basic requirements of the profession of chemical engineer remained very vague.

But then there arose an ever more urgent need for specialists who could scale up the work of the laboratory to an industrial level in optimal conditions of safety and productivity. As the chemical industry developed, particularly in Germany and in the United States, more complex chemicals were produced in large quantity, and firms required technicians who were better trained than in the past, and who, among other things, had mastered the language of chemistry codified in the periodic table of the elements established by the Russian chemist D. I. Mendeleyev in 1869.

Companies in the United States were not yet looking for high-powered specialists, but only for technicians sufficiently versatile and well trained to develop methods of production, to follow its different stages, and to make the connection between laboratory experimentation and large-scale manufacturing. To do this, companies employed "industrial chemists" who were graduates of American, and in some cases German, universities. Between 1861 and 1899, 220 doctorates in chemistry were awarded by six universities, a figure that rose to several dozen per year between 1900 and 1914 (97 in 1914). Censuses counted about 9,000 chemists in 1900 and 16,000 in 1910 (some 500 with doctorates); the vast majority of these were employed in industry.[32] To this number must be added the many mechanical engineers working in the chemical industry who, though often assigned to the construction of equipment, could also—provided their knowledge of chemistry was reasonably adequate—work on the development and implementation of new production processes.

In 1888, Lewis Mill Norton, a professor of industrial and organic chemistry at MIT, at the time still a fairly obscure technical institute, created a four-year bachelor program entitled "Course X" (i.e., ten), "arranged to meet the needs of students who desire a general training in mechanical engineering and to devote a portion of their time to the study of the application of chemistry to the arts, especially to those engineering problems which relate to the use of and manufacture of chemical products." Norton had only one assistant and one instructor.[33] The course guide indicates that courses in chemical engineering were quite close to those in mechanical engineering, the only addition being chemistry. Norton's aim was to give MIT graduates a leg up in the labor market by virtue of their comprehensive training. The same year, the president of MIT, anxious to distinguish it from other universities and to make it better known, lauded the creation of this course and explained in his annual report that

> the chemical engineer is not primarily a chemist but a mechanical engineer . . . who has given special attention to the problems of chemical manufacture. There are a great number of industries which require constructions, for specific chemical operations, which can best be built, or can only be built, by engineers having a knowledge of the chemical processes involved. . . . Heretofore, the required constructions have, generally speaking, been designed, and work upon them has been supervised and conducted, either by chemists, having an inadequate knowledge of engineering principles and unfamiliar with engineering, or even building practice; or else by engineers, whose designs were certain to be either more laborious and expensive than was necessary or less efficient than was desirable, because they did not thoroughly understand the objects in view, having no familiarity, or little familiarity, with the chemical conditions under which the processes of manufacture concerned must be carried on. It was to meet this demand for engineers having a good knowledge of general and applied chemistry, that the course in chemical engineering was established.[34]

It was not long before other universities followed in the footsteps of MIT and created their own courses in industrial chemistry. This was done at the University of Pennsylvania in 1892, at Tulane University in New Orleans in 1894, and at Michigan and Wisconsin Universities in 1898. Still essentially courses in mechanical engineering, they were more numerous at some institutions than at others. Sometimes they were offered by chemistry departments (in divisions of "applied chemistry") or, more rarely, by departments of mechanical and even electrical engineering.[35] At the University of Pennsylvania, for example, the program in chemical engineering for the year 1892–93 (for third-year students) consisted of two courses in chemistry (analytic and organic chemistry), two laboratory courses in physics, one course in metallurgy, four in mechanical engineering (electricity, hydraulics, steam, and a laboratory) and one in calculus.[36] The graduates easily found employment in one of the numerous chemical firms in Philadelphia and the surrounding area.

The chemical engineers trained in the United States before about 1920 were still close cousins to the mechanical engineers, even though by 1905, the courses in industrial chemistry offered at MIT had begun to place greater emphasis on chemistry at the prompting of William H. Walker, a young professor. Yet the difference in the chemical engineer's identity was not very obvious, and no one knew quite how to define a profession that was still in limbo. By contrast, electrical engineering, though also originating in departments of mechanical engineering, had more quickly found its own identity, because it was structured by an electrical industry concentrated in a few giant companies, by a body of knowledge and practices that had rapidly stabilized, and by the absence of competition from the physicists.

This "mechanical" cast of the first chemical engineers may be explained by the fact that the American chemical industry of the time was still largely oriented toward heavy inorganic chemistry and the production of heavy chemicals such as sulfuric acid and soda, which called for a more mechanical than chemical approach to manufacturing problems. Making a well-known product like sulfuric acid did not require highly sophisticated chemical analyses. Nonetheless, in order to be profitable, it had to be manufactured in very large quantities, and this involved a certain number of mechanical problems. Yet at this same time, German industry placed major emphasis on scientific research and on sophisticated organic chemistry, employing large numbers of high-powered chemists, particularly in the effort to develop its famous synthetic dyes.

The American chemical engineers—or perhaps one should call them proto–chemical engineers—were above all concerned with issues of quantity, yield, and continuous-flow production. Their attention was focused on machines and on the proper organization of production, inspired by Frederick W. Taylor's concept of "scientific management." At the end of the century, Taylor was working hard to propagate his famous scientific organization of work.[37] His business card identified him as a "consulting engineer. Systematizing Shop Management and Manufacturing Costs a Specialty." By this time, he was surrounded by disciples, the "Taylorites," who endeavored to improve and refine the master's principles, which spread throughout American industry, including the chemical industry—no doubt by way of the electrochemical industry, where metallurgists and chemists could exchange ideas.[38] We shall see further on how DuPont made an effort to reorganize its plants in keeping with Taylorite principles. For the moment, I simply wish to stress that the proto–chemical engineers were not yet a profession with a codified and socially valued body of know-how. They were viewed rather as poor relations of the mechanical engineers, who by then were at the height of their renown; some of them were actually devising grandiose projects of social engineering.[39]

By the late nineteenth century, DuPont was already hiring large numbers of chemists and engineers in different specialties. Among them, there were indeed some chemical engineers, but their careers were not particularly brilliant. As the sparse data available for the period before World War I indicate, these engineers tended to be confined to rather marginal tasks, and their career prospects were relatively limited.[40] By the same token, other engineers with recognized and more prestigious training were monopolizing the mid- and upper-level positions in the company.

Searching for DuPont Engineers

To begin with, let us look at the powder mills, the locus of DuPont's traditional know-how. The manufacture of black powder had hardly changed since the early nineteenth century, except that sodium nitrate had replaced saltpeter. Once the nitrate was mixed with charcoal and sulfur by means of millstones, it was compressed and dried. The gunpowder was then stored in wooden barrels. These operations were carried out in small workshops, well separated from each other along the Brandywine River, so that an explosion would not endanger the entire site. The workers, veteran powder makers who had been with the company all their working lives—most were from Ireland or Italy, but some were of French origin—knew better than anyone how to avoid catastrophic accidents. They used iron-free tools, prudently moved away from the shop in thunderstorms, and handled their material without excessive haste. Nonetheless, theirs was a dangerous occupation, and they derived a certain pride from it, not unlike miners in deep mines: daily exposure to danger contributed to a specific kind of working-class culture, which was to some extent cut off from its equivalent in the heavy iron and steel plants and other metal works. Powder mills were located in isolated places, and the company liked to hire the sons of its employees, who had been familiar with the job since early childhood.[41] Making gunpowder and explosives required reconciling sometimes contradictory imperatives: it was necessary to produce in large quantities, particularly in times of war; yet this had to be done under acceptable conditions of safety. These two aspects governed a gamut of technical practices developed through long experience, in which the workers' know-how continued to be of prime importance.

Black powder manufacturing was the archetypal heir to a workshop culture characterized by a semi-artisanal approach. What it required above all were rigorous safety procedures; no profound knowledge of chemistry was necessary. Its scientific and technological aspects were marginal.[42] One powderman said that "there are many things [in gunpowder making] that can't be told. They must be learned by experience."[43] There was not much to do for chemical engineers in such a context, where manufacturing processes were well established and entrusted to foremen and experienced workers. By contrast, DuPont did employ mining and civil engineers to oversee blasting projects in mines and on construction sites (for canals or railroad tunnels).

The production of black powder declined steadily, however, until it came almost to a complete stop in the 1950s. By the beginning of the twentieth century, new explosives derived from nitric acid—dynamite and smokeless gunpowder—had begun to conquer the military and civilian markets.

The explosives industry originated with Alfred Nobel's invention of dynamite in 1865. Dynamite, which consists of a mixture of stabilizing agents and nitroglycerine, allowed for the safe use of the latter product. Invented in 1846 by the Italian chemist Ascanio Sobrero, pure nitroglycerine had been abandoned because of the disastrous accidents it had caused. In a parallel development in the last years of the nineteenth century, the military chose smokeless gunpowder over competing products to propel shells out of cannon. Smokeless gunpowder is made from cotton or wood pulp impregnated with nitric and sulfuric acids, along with different stabilizing agents, whereupon it is compressed, dried, and sold in the form of grains of various calibers. The first smokeless gunpowder for military use was the famous *poudre B* invented by the French engineer Paul Vieille in 1886 and named in honor of General Georges Boulanger, the French minister of war and would-be dictator. Other countries soon followed suit, among them the United States, which adopted smokeless gunpowder in 1900. Dynamite and smokeless gunpowder are chemically cousins.

The manufacturers of explosives, and especially DuPont, faced two main problems: one was the supply of raw materials for making nitric acid, the essential ingredient of the explosives (and we shall see presently that it took a revolution in the chemical industry to deal with the issue of nitric acid); the other was the elaboration of manufacturing processes that reconciled safety concerns with mass production.

As it happened, the reorganization and expansion of their chemical company provided the DuPont leadership with the context in which to attempt a reorganization of the different production units along Taylorist lines on the eve of World War I, between 1912 and 1914.[44]

In the early years of the twentieth century, Taylor's growing reputation had not failed to attract the attention of DuPont's directors. T. Coleman du Pont actually knew Taylor personally, having made use of his services in 1896 for setting up a system of cost control at the Johnson and Lorain Steel Company he was running at the time in Lorain, Ohio.[45] In 1911, the year when the famous *Principles of Scientific Management* came out and when Taylor's renown was at its height, Hamilton Barksdale, vice president of DuPont, wrote to Taylor to ask for copies of his book and to tell him how interested he was in his work.[46] In the months that followed, an "efficiency division" was set up in the Explosives Department. Yet the very notion of efficiency was a problem in the explosives industry, for it could not be reconciled with that of productivity.

In keeping with the principles of scientific management, the rationalization of production was carried out in several stages. First of all, one had to find the best way to perform an operation by carefully observing a production process; secondly, the principles of that process had to be standardized; and then the workers had to be trained in this new method and motivated to follow it by means of raises and bonuses. According to Taylor, an industrial process is efficient when it allows for the highest possible increases in productivity. The famous engineer could not be bothered with psychological or cultural considerations, any more than with issues of safety, which he practically does not mention at all. His system envisioned efficiency in the short term, without factoring in the consent of the labor force, its health, or its safety.

In the gunpowder and explosives industry, however, operations carried out too hastily could lead to an explosion, an exceedingly costly occurrence in human and material terms. The speed of production had to be subordinated to the imperative need for safety. This was recognized by the efficiency division as soon as it was set up, for its head noted that it would not be appropriate to offer the financial incentives that Taylor considered so important for increasing the workers' productivity, and that such rewards had to be related to the quality of their work and their attention to safety rules.[47] Moreover, the evaluation of individual merit was difficult in a context of team work and interchangeable skills, which also involved safety considerations.

Nevertheless, Barksdale decided to reorganize some workshops at the Repauno dynamite plant according to Taylorite principles, albeit on a trial basis. This required, in particular, that the workers agree to obey the foreman in charge of the workshop reorganization in all things. This attempt ended in failure. On 22 September and on 8 December 1913, two explosions killed four and six workers respectively. The press reported these events in detail, indicating that the workers had been rushed by the foremen, and that productivity had been increased to an unreasonable level, in short, that the new methods had been responsible for the accidents.[48] It appears that the company's directors shared this judgment, for the efficiency division was quickly and discreetly dismantled shortly thereafter.

This inauspicious experimentation is interesting in more than one respect, for it makes us understand that in these first decades of the century, the famous "modernization" of American industry was not necessarily synonymous with Taylorization: there was no "Taylorite turning point" in the gunpowder and explosives industry. Here the continuous-flow processes that became increasingly common in the metallurgy industry and in the manufacture of certain inorganic chemicals such as sulfuric acid, were rejected for safety reasons. Preference was given to processes carried out batch by batch, in isolation from one another. Not that the DuPont managers ignored Taylor; quite the reverse. But having examined his principles, they came to the conclusion that a prudent and incremental modification of tried and true processes under the supervision of safety experts working in a safety division created in 1916 was more satisfactory.

In this context, war broke out in Europe. Pierre S. du Pont's biographers demonstrate how DuPont was able to fill the massive orders placed by the Allies, and particularly the French (who as early as October 1914 ordered 4,000 tons of military gunpowder) by concentrating on financial and organizational problems. Some weapons manufacturers, such as Winchester and Remington, unable to solve these problems, went bankrupt due to the combination of costly investments in increased production capabilities and delayed payments from their European customers.[49]

Except for the delicate operations of processing methylbenzene (toluene) and phenol, which were carried out by chemists, the manufacturing of smokeless powder was entrusted to experienced technicians, as well as to mechanical and civil engineers. The rapid construction during the war of smokeless powder factories, such as Old Hickory in Tennessee and Hopewell in Virginia, and also TNT and picric acid plants (at Deepwater, New Jersey), was a tour de force that astounded the contemporaries, but there was really nothing new about these large-scale production operations. The chemists essentially had the task of monitoring the quality of the products and ensuring the safety of these processes (it was essential, in particular, to verify the exact power and velocity of the explosives.) The production of gunpowder and explosives, then, is part of the history of mass production and represents one of its original facets. Until World War II, the explosives industry remained a heavy industry that employed tried and true processes it had inherited from the traditional powder mills.

Engineers at DuPont

There were chemical engineers in DuPont's Engineering Department, which had grown out of the Engineering Division of the Department of High Explosives, founded in 1902 to design and build the company's plants and industrial equipment. A year later, the Light, Heat and Power Division was established, still within the Department of High Explosives. The Black Powder Department also had a Construction Division.[50] The war and the construction of gunpowder and explosives plants brought about the merging of these divisions and the creation in 1915 of a legally independent company, the DuPont Engineering Company, which returned to the mother house in 1928 under the name of Department of Engineering.

A civil engineer who bore the title of chief engineer always headed this department. Of the twenty-seven engineers in it who can be traced, seventeen were civil engineers, four were mechanical engineers, one was a mining engineer, and five did not have college degrees.[51] Some of them were former military engineers. Quite a few later went to work for General Motors, for which the Engineering Department had already built several plants in the early 1920s, among them the Cadillac and Buick plants at Detroit and Saint Louis. As for the Department of Chemical Research, it was headed by Charles Reese, a graduate of the Heidelberg and Johns Hopkins Universities, who worked with a small team of chemists that did not include any chemical engineers.

Chemical engineers did work in explosives plants before World War I. In this respect, the early stages of Robert MacMullin's professional career are revealing.[52] MacMullin, born into a New York lower-middle class family in 1898, entered MIT in 1916. His freshman year went well enough, but when the United States joined in the war, young MacMullin chose to interrupt his studies for both financial and professional reasons: he wanted to find out what precisely was involved in the work of a chemical engineer and also to earn his living in order to make things easier for his family. The young man was granted an interview by Charles Reese, the director of research at DuPont, and was hired on the spot, even though at that point, he only had one year of college. DuPont was short of personnel at the time, and Reese was not the man to pass up a resourceful young fellow and a good chemist.

The firm sent MacMullin to the Deepwater plant on the banks of the Delaware River, where DuPont was engaged, among other things, in efforts to synthesize indigo, and where it produced TNT and the nitrite-based picric acid that the French artillery needed in large quantities for the shells of its 75mm cannon. Living in Wilmington, the young man had to take a steam ferry every day to make his way up the Delaware to Deepwater. Once he had arrived, he changed his clothes, donning, like all the other employees, woolen overalls and shoes with bronze nails in order to avoid causing sparks. In addition, guards checked for tobacco and matches. And woe to the careless! When he walked into the plant for the first time, MacMullin choked on the strong acid vapors emanating from the vats of sulfuric and nitric acid.[53] The foremen who offered him chewing tobacco had teeth "corroded by the acid down to the roots." The indigo workers proudly showed him their skin, dyed blue by the aniline.[54]

The new recruit was assigned different tasks, some of them dangerous, in the improvement and the maintenance of different sections of the plant. At one point, he had to slide under a vat containing a toxic product (toluidine) and was sprayed with it: at Deepwater, accidents happened frequently, and deaths were not rare.

MacMullin, who was paid $90 a week, had a position as "apprentice chemical engineer," but there is no doubt that what he needed in his work was not so much a thorough knowledge of chemistry as practical know-how and a good bit of resourcefulness. Reese was pleased with him and would have liked to keep him, but MacMullin soon left DuPont to join the Army, for he wanted to go overseas and fight before the end of the hostilities.[55]

It would seem natural for a chemical company to place a chemical engineer in an eminent position. But then technical functions do not automatically spring from technical change, as if professional delineations were imposed by an inflexible logic of production—as if, indeed, technology were a modality unrelated to society. The different fields of engineering were the products of social arbitration—of which more will be said later—rather than of an abstract modernization that would conveniently have dictated their launching and their characteristics. The American chemical industry could have come into being and evolved without chemical engineers, through other technical, social, and professional arrangements. And in fact, until World War I, industrial leaders were probably quite satisfied with the division of labor among chemists, civil and mechanical engineers, and chemical production engineers.

Beginning with the new century and the arrival of the du Pont cousins, their firm became one of the great American companies at the cutting edge of new management techniques. Managers with academic degrees were hired to take the place of members of the du Pont family, separate departments with precisely defined tasks were created, and rigorous accounting methods were instituted, along with long-range planning of the firm's activities.[56]

The mainstay of DuPont's modernity in management were these managers, who had a highly technical approach to their work. In this respect, DuPont was different from a company like Ford, which valued practical engineers, men who did not hesitate to crawl under a machine tool in order to repair it, but who were most uncomfortable when it came to management—paper-shoveling they called it.[57]

Examining the backgrounds of the group of managing engineers who set up DuPont's multidivisional structure in 1921 is quite revealing in this respect. Hamilton Barksdale, a close associate of the du Ponts, had graduated from the University of Virginia with a degree in civil engineering. He was hired by DuPont in 1892, almost by chance, on the occasion of a stay in the Wilmington area, when he was working for a railroad company. In 1893, having married the daughter of Victor du Pont, Barksdale became general manager of the Repauno plant, and then one of the company's vice presidents in 1902, when the du Pont cousins took over. Barksdale was an extremely able manager, and he trained a group of young engineers in this field. Its principal figures were Frank Mc-Gregor, an electrical engineer from MIT, Harry Haskell, a mining engineer, Bill Spruance, a graduate in electrical engineering of Princeton, John Pratt, a civil engineer and graduate of Virginia like Barksdale, and F. Donaldson Brown, an electrical engineer with degrees from Virginia Tech and Cornell—who eventually married Barksdale's daughter.[58] None of these young elite managers was a chemical engineer by training, and indeed none of them knew much about chemistry. Barksdale knew nothing about it, as he freely admitted.[59] The first managers at DuPont were not, strictly speaking, specialists in their field, but rather generalists in management who worked for DuPont as they would have worked for any other large company. Actually, some of them, such as Brown, left DuPont and exercised their competence on behalf of General Motors when that firm was reorganized by the du Pont family, which had gained control of it during the World War I.[60] And when, at the end of the war, Pierre S. du Pont felt that the careers of a certain number of these engineers should be speeded up in preparation for the takeover, the Executive Committee opened its doors to six new managers, none of whom was a chemical engineer.[61]

Professional Status and Representations

At a more general level, conflicts of interest and disagreements over professional boundaries also played a part, as the story of the American Institute of Chemical Engineers (AIChE), the professional organization founded by a few chemical engineers in 1908, demonstrates.[62] Chemists, many of whom worked as industrial chemists, looked askance at the intrusion of these new engineers. Many of them feared that it would affect their professional status, and this concern was particularly strong among the production chemists, who worked outside the laboratory on the design, construction, and running of chemical plants, for the newcomers directly threatened their functions.

Thus the powerful professional association of American chemists, the American Chemical Society, opposed the creation of the AIChE. Its leaders made the general point that no new discipline was needed to design chemical plants. Following the German model in this instance (for many American chemists knew it well, having studied in Germany for a time), they pleaded for maintaining the traditional distinction between chemists and engineers. This did not mean that they could not closely work together; however, the blurring of boundaries embodied by the chemical engineer should be avoided. But, of course, the AIChE was explicitly created for the purpose of giving an institutional identity to chemical engineers and encouraging industrialists to hire them.[63]

The mechanical engineers for their part also wished to maintain the status quo of their discipline. The historian Terry Reynolds reports that Charles Lucke, professor of mechanical engineering at Columbia University, explained that since the equipment found in chemical plants was, after all, basically the same as in other industries, mechanical engineers were best suited to engage in fruitful dialogue with chemists.[64] Both the production chemists and the mechanical engineers feared that fewer of them would be hired by chemical companies, and that they would be supplanted by chemical engineers.

But perhaps something else that was more essential was involved here. For what was taking shape along with chemical engineering—not yet very clearly before the 1920s, but it will do as a temporary marker—was a general reorganization of the chemical industry that would demand complex processes of exchange and negotiation between different institutions. Its development would have far-reaching consequences for the organization of the chemical industry and its relations with its outside partners.

In the first decade of the twentieth century, the profession of chemical engineer was statutorily recognized, thanks to university courses and the AIChE, whose stringent criteria for admission expressly stipulated expertise "in chemistry and in mechanical engineering applied to problems of chemistry." To become a full member of the association, one had to be at least thirty years old and to have had a minimum of ten years' professional experience, if one did not have an academic degree, and five years' if one did. The emphasis was placed on experience rather than on academic qualifications, unlike in the association of electrical engineers, which required graduate studies of its members. But the association attracted few members: 40 in 1908, 344 in 1920, 644 in 1924, and 872 in 1930.[65]

The AIChE's efforts to establish professional boundaries did not create a discipline; they only circumscribed an area of competence, without forging a clear-cut conceptual identity. Hence the importance the AIChE attached from the very beginning to the training of chemical engineers, unlike the other engineering associations, which, especially at that time, were most interested in their social and technical functions. It should be added that before 1919, the AIChE's *Transactions,* its official publication, featured little more than meticulous descriptions of chemicals, the laborious inventories of a discipline that remained to be invented, so that one understands why other engineers looked at it with some condescension. The chemical engineers still had a long way to go before they made their mark at DuPont, particularly in its upper echelons, among the men who had just moved into a large, brand-new building that proudly rose in the center of Wilmington, Delaware.

Even before World War I, the DuPont directors had begun to diversify their firm's production—although gunpowder and explosives still accounted for more than 90 percent of its sales in 1913. This diversification took place in the area of cellulose chemistry, a field deriving from that of the explosives that DuPont had been manufacturing for several decades. However, the classic and proven manufacturing techniques for gunpowder and explosives were not suitable for making imitation leather, fabrikoid, and other cellulose products the company was now beginning to produce. To be sure, DuPont had gone about this by acquiring other firms, and its directors had placed some hope in the know-how of the personnel of the Fabrikoid Company, which it had acquired in 1910; the Arlington Company (celluloid), acquired in 1925, the Fairfield Company (plastic coating), bought in 1916; and then a series of paint and varnish companies (Harrison Brothers, Becton Chemicals, Flint Varnish, Chicago Varnish, and the like).[66] But the acquisition of these firms proved to be a disappointment. Fin Sparre, the director of the Department of Development, was chagrined to report that the practices used by Fabrikoid were archaic and actually dangerous.[67] The purchase of Arlington was disappointing as well. In order to carry out the reorganization of these companies and pursue their own strategy of diversification, the DuPont directors felt that more attention than previously should be paid to methods of production, which should be governed by principles of safety, economy, and simplicity.

Advised in these matters by its "strong arm," the Department of Development, DuPont's Executive Committee decided to consolidate its geographically dispersed companies in order to achieve substantial economies of scale. This called for improvements in management—unifying the sources of supply of raw materials and the methods of recruiting management and labor—as well as synthesizing methods of production. In this endeavor, the stage leading from laboratory experimentation to industrial production was of crucial importance. The more sophisticated chemical research becomes, the more difficult it is to scale up to the level of mass production. This stage of scaling up was thus not a matter of course, for the chemist's experimentation was not necessarily understood by the men who ran the plants, be they chemical or mechanical engineers. "I can well remember," Charles Stine—at the time director of chemical research at DuPont—testified in 1928, "when the practical man and the scientifically trained man looked askance at one another, the one alleging that the practical man knew nothing and worked entirely by rule-of-thumb; the other that the laboratorian was so foolishly impractical in his ideas. . . as to incur, at best, nothing more than a tolerant commiseration."[68]

By the end the 1910s, criticism of the chemists in DuPont's Department of Research mounted. Observers took them to task for their inability to translate the work of the laboratory to the industrial level and for their lack of knowledge about production and markets. This was particularly true in the field of dyestuffs, which was new for DuPont and still presented enormous difficulties of research and manufacturing. On the basis of incorrect and imprecise data, and at a cost of $3 million, DuPont had built and then had to rebuild one of its dye plants. Chemists complained that plants often went up before the laboratory work was completed.[69] Reese and his associates wished to keep control of all the company's research and development activities, while many of the managers wanted research and development activities to be decentralized and better coordinated with production. The same considerations were advanced by the Engineering Department, except that the latter could point to the importance of its war-related work when it called for the building of giant explosives plants. Fletcher Holmes, director of the Laboratory of Organic Chemistry, stressed the need for building pilot plants and accumulating data before initiating the industrial stage.

Out of this questioning of the firm's organization came the famous decentralization of 1921, which also affected the Department of Chemical Research. What was most notably involved here was the juncture between research and industrial production—and also between production and marketing. The response of the DuPont directors was actually twofold: they opted for decentralization in order to bring the researchers closer to the plant managers, but also for the training and recruitment of men capable of serving as liaisons between the laboratory and the plant.[70]

However, chemical engineers did not yet occupy a position in which they could serve as the intermediaries between theory and practice of whom the firm had increasingly urgent need. The strategy of diversification brought to light two major difficulties: on the one hand, the question of the firm's structure, and, on the other, that of the human resources available to carry it out. However clear the overall DuPont strategy, function did not automatically create managers or engineers. Diversification and its organizational consequences took place at a time when the professional element capable of putting it into practice was not available. DuPont adopted an unrealized or, more precisely, pending strategy, a "virtual strategy."[71]

Moreover, the indeterminate status of the profession of chemical engineer before World War I brings up the larger issue, investigated by Luc Boltanski, of how to establish professional categories. In France, the category of *cadre* (midlevel management), originally a statistical construct, became a reality thanks to those concerned themselves, who turned it into a social group combining disparate elements of the bourgeoisie and petite bourgeoisie.[72] In the United States, there is no equivalent of the French term *cadre*: here, one deals either with engineers or with managers, or, more generally speaking, with white-collar workers. However, the term *engineer* is not as strictly defined in the United States as it is in France, where there is a central oversight authority (the Commission des Titres d'ingénieur), and where the state has played a major role in the training and the recruitment of engineers. The fluidity of professional status in America was even more pronounced in the case of the early chemical engineers, for despite the stringent criteria for joining the AIChE, their job was not considered a profession: the function had been established, but no codified training (and hence no specialized body of knowledge) had yet been put in place; nor was this an aggregate comparable to the French *cadres*. This made for imprecise definitions: when evoking his employment at DuPont, MacMullin indiscriminately speaks of himself as an "engineer" or a "supervisor."[73] It is also significant that the AIChE was not a member of the national engineering societies. In 1920, one of its members complained that "we are not able to persuade the other engineers that we are indeed engineers."[74]

As far as statistics are concerned, the Census Bureau did not list chemical engineers as an autonomous category until 1940 (when it counted 13,000 of them).[75] Before that date, they were officially included in the category "metallurgists and metallurgist engineers" (14,000 in 1930), but some of them surely appeared in that of "chemists" (45,000 in 1930), on the basis of the declarations of these individuals themselves, which in turn were based on academic degrees, functions, and subjective perceptions of their work and their professional status. This statistical recognition came thirty years after that of the electrical engineers (whom the census distinguished from the mechanical engineers as early as 1910)—another indication of the low social visibility of the chemical engineers until the end of the 1930s. The census of 1940 confirmed, and indeed consolidated, the professional and social emergence of these engineers. It took notice of the increasing coherence of a professional group that would henceforth have considerable visibility.

We should therefore not try to give an overly strict definition of this nascent profession. For if we did, we would miss the social efforts that shaped this group and suggest that chemical engineers have always existed (after all, even in antiquity "people did chemical engineering without knowing it"), in other words, that the function exists ahistorically, and that technology is mechanically reflected in the professional structure. Rather than trying to zero in on a set of objective criteria that would infallibly allow us to find the chemical engineer of the first two decades of the twentieth century, we must bring to light the social, professional, and technological process that was to give him his own identity in the following decades. Clearly, the early performative step of creating a professional association was insufficient to establish this solid identity.

Nor was this achieved by a hypothetical decision—it does not appear in the archives—of DuPont's Board of Directors to bet strategically on the chemical engineers and to reorganize the firm accordingly. To believe this would be to reinterpret the history of the firm a posteriori and to confer upon it a linear and progressivist logic flowing from the allegedly rational and visionary choices made by its leadership. In reality, it appears that the demand for chemical engineers followed the reorganization of 1921, when diversification shed a stark light on the detrimental lack of specialists in industrial methods. The professional advancement of the chemical engineers was a consequence of the reorganization, rather than the fruit of an explicit strategy. The fact is that a new cohort of chemical engineers appeared on the labor market at that precise moment. This was not a coincidence, insofar as these engineers of a new kind were trained by universities attuned to the demands of industrialists in the chemical field. Despite the advantages of hindsight, one must resist the temptation to construct a functionalist model that integrates the various institutionalist strategies in too harmonious and convergent a manner. New techniques of production and organization could sometimes win adoption simply in order to deal immediately with specific difficulties, without anyone's thinking in advance of all their consequences.[76] This observation mitigates the functionalism that always assumes the existence of a rationalizing strategy beneath technical and organizational choices.

Beginning in the early 1920s, chemical engineering experienced a spectacular development, both in quantity and in quality. What were the methodological and institutional stages of this rapid rise?

Inventing a Discipline: Unit Operations and World War I

A new kind of training in chemical engineering was instituted in the early 1920s, first at MIT and then at most other American universities. Based on a new approach to the production problems of industrial chemistry, it was more methodical than earlier programs, centered on the study of processes rather than of products, and the solutions it offered were adapted to the needs of the chemical industry and its academic partners. This approach was comparable to Fordism, whose principles rapidly spread throughout the major American industries from around the time of World War I.

The conceptual foundation on which the reorganization of work was based was that of unit operations. This conceptual foundation is important, for it became one of the essential stages in the development that transformed American chemical engineering from a mechanic's know-how to a more theory-based approach to the processes of production, which for a long time distinguished it from the practices of industrial chemistry in Europe.

The year 1915 is a convenient landmark in the history of American chemical engineering, for it was then that the consulting engineer Arthur D. Little wrote a report for the president of MIT in which he proposed restructuring the courses in industrial chemistry around the concept of unit operations, which was here formulated for the first time.[77] At the time, Little was chairman of the visiting committee for MIT's Department of Chemistry and Chemical Engineering, while also directing a respected firm of consulting engineers, Arthur D. Little Inc., which he had founded with his friend William Walker in 1902. Little hoped that the courses at MIT would become the model for the training of all future chemical engineers throughout the country. His report was therefore intended to reach beyond MIT and to influence the academic policies of the other universities. It was greeted with enthusiasm by the Association of Chemical Engineers.[78]

A unit operation is a basic operation in industrial chemistry—crystallization, distillation, combustion, filtering, heating—that makes it possible to break down a process of production into a series of simple and standardized operations.

The unit operation makes it possible to express the different manufacturing processes in a synthetic and concise manner: instead of being described like cooking recipes, they can be analytically transcribed by structuring them around standardized operations. At a time when the chemical industry was developing rapidly, exhaustive descriptions were no longer possible. What would be the use, for instance, of teaching the minutiae of making caustic soda to people who would spend their careers making petrochemicals? And yet these two activities present analogous problems, such as those related to the evaporation of solvents. This being the case, it is wiser to analyze the principle of evaporation synthetically, independently of particular applications. In this manner chemical engineering could be reorganized by studying processes rather than products.

The textbooks of chemical engineering and the specialized journals changed markedly from the early 1920s on, when the traditional designations gave way to synthetic information and standardized terminologies.[79] As Little wrote, chemical engineering is not "a composite of chemistry and mechanical and civil engineering, but a complete specialty of its own, based on unit operations. The latter, when properly coordinated, constitute a chemical process on an industrial scale."[80] What, then, were the concrete changes brought to industry by the unit operations?

Until the 1920s, chemical plants were built without preconceived plans. Once the plant was finished, it took a fairly long time to "break it in," that is, to make the adjustments, additions, and improvements needed to achieve the desired results. Industrial processes were codified in a rather rudimentary fashion, and in a sense, every plant was a unique case and their methods were hardly standardized. At DuPont, the principal element of standardization was contained in the blueprints of the Department of Engineering and in the logbooks of the production engineers and foremen. The production supervisors used these logbooks to record the timing of operations, the quantities produced, problems encountered, and a great deal of other information concerning the methods used. They then made these records available to their colleagues, so that a body of practical know-how was passed around on a daily basis at the production sites. The logbook was an intermediary link between codified instructions and blueprints and the oral transmission of experiences. MacMullin mentions the logbook that he "personally" handed over to the engineer of the night shift. If the engineer failed to consult the logbook right away, he could get into trouble, as did a careless colleague of MacMullin's who noticed a breach of the safety rules only long after the problem had been disclosed in the logbook.[81] The log was precious, and people used their best handwriting in it and were careful never to mislay it. Early in the century, DuPont built a plant for producing soda by electrolysis. An engineer who had come from Europe knew about this process, "and he taught the others how to run the factory, without theory."[82] His knowledge spread throughout the firm by means of the logbooks, from which everyone could benefit, even after this particular engineer had left DuPont for Union Carbide.

What did the DuPont plants look like in the first decade of the century? There were two major types of plants. In those that produced inorganic chemicals by means of traditional processes, continuous-flow processes had been put into place. This was how sulfuric acid was produced in very large quantities. But there were others that produced light chemicals and explosives in batches. The batch process consisted of carrying out a series of chemical operations for a limited quantity of product, and then repeating the various stages for another batch.[83] This generally more artisanal method was in use in organic

chemistry workshops, where cellulose, dyestuffs, and various acids were manufactured. In order to obtain nitrocellulose, for instance, an operator mixed the
different ingredients (toluene, nitric acid, cellulose) in a vat, carefully watching
the temperature as he did so; if it rose above a certain level, he had to open
a safety spigot, through which the mixture would flow into an outside basin
filled with water, where it could cool. Subsequently, the operator could recover
the mixture by making use of the differential between its density and that of
the water. Unit operations, as defined by Little and elaborated in *Principles
of Chemical Engineering*, a hefty textbook that three MIT professors, Warren
Lewis, William Walker, and William McAdams, published in 1923, made it possible to improve the industrial processes and facilitated the changeover to continuous-flow production.[84]

Principles of Chemical Engineering supplied a discourse on method for the
discipline. The engineer was called upon, in a first phase, to *recognize* the questions, that is to say, to break down the process into simple operations, such
as heating a solution (first operation), followed by separating the gases (second operation), and then condensing the gases (third operation). Thereafter,
every unit operation had to be *treated* individually (for instance, by means of
Bernouilli's equations in the case of separating gases). The necessary equations
appeared in detail in *Principles of Chemical Engineering*. The engineer thus had
at his disposal a maximum of data before proceeding to the construction of a
plant, an advantage that in principle allowed it to become operational more
rapidly. Unit operations established principles common to the entire chemical
industry, since an operation like condensation, for example, is involved in a
great many industrial processes. The roughly 500 pages of *Principles of Chemical Engineering* justified a new approach to the problems of production: all the
engineer had to do was to break them down into a sequence of the most basic
operations possible.

And, finally, the advent of unit operations required the invention of a language and the systematizing of a body of know-how. Unit operations spelled
the end of pure empiricism and of learning by doing; they standardized production and the exchange of know-how, and they established rules for hitherto
empirical techniques. In the realm of practical work, they endeavored to set
up one way of doing things. In this respect, the birth of chemical engineering cannot be dissociated from scientific management and from the American
manufacturing system. The search for the smallest common denominator in
unit operations was in fact inspired by the Fordist principles that were just

then spreading throughout American industry.[85] As a consulting engineer, Arthur Little was well acquainted with the Ford plants at Highland Park, and the breakdown of the assembly of the Model T Ford into a series of elementary operations had certainly captured his attention. Unlike the DuPont plants, the Ford plants could easily be visited, and American and European engineers came to see them in droves.

In a sense, Little was to chemical engineering what Ford was to the metal industry. Ford's celebrated article "Mass Production" in the *Encyclopedia Britannica* asserted: "Mass production is not merely quantity production, for this may be had with none of the requisites of mass production. Nor is it merely machine production, which also may exist without any resemblance to mass production. Mass production is the focussing upon a manufacturing project of the principles of power, accuracy, economy, system, continuity, and speed."[86]

Even if the famous saying of the Bostonian businessman Edward Filene, "Fordize or fail," in the end turned out to be wrong—since mass production, as it developed in the United States, largely freed itself from the Fordist precepts—it is true that the idea of mass production for the mass market spread rapidly and widely, in both the chemical and metallurgical industries.[87] By substituting charts and graphs for the series, and structure for the catalogue, Little and his unit operations were about to thrust the chemical industry into the new century.

This does not mean, of course, that the unit operations completely changed the chemical industry overnight. Nonetheless, they did propose a new language of modernity, rhetorically structured around the notions of efficiency, rationality, and output, that was designed to give the chemical industry and its technicians visibility in the new era. Walker himself was a progressive modernizer and thought along similar lines to Herbert Hoover. To quote Cecilia Tichi, the chemical industry "shifted gears" when it adopted a favorable-sounding modern language, whose performative effects were by no means negligible either.[88] The rhetoric of change and its reality were inextricably linked.

How Innovative? How Exceptional?

Owing, no doubt, to the influence of metallurgy, the chemical processes outlined in *Principles of Chemical Engineering* were seen in a very mechanical manner. Most of the unit operations, such as "grinding" and "pulverizing," were directly taken over from mechanical engineering, but others, such as "heat transfers," were already informed by a more mathematicized and scientifically more advanced culture. By the end of the first decade of the twentieth century, unit operations represented a transitional stage: while still imprinted with the principles of mechanics, they already offered the possibility of a theoretical and mathematical approach to the construction of the means of production.

As always when it comes to innovative processes, the question arises of how much of a rupture introduction of the new concept caused. According to the historian Martha Trescott, unit operations did no more than formulate practices that already existed in the chemical industry. For instance, the English engineer George Davis, professor at the Manchester Technical School, in his *Handbook of Chemical Engineering* (1901), had already broken down the operations of industrial chemistry into standardized physical operations. It is also true that at the turn of the century, American industry produced certain chemicals in large quantities by applying the principles of continuous flow and Taylorist management. There is no absolute beginning in history. Think of the introduction of the assembly line by Ford in 1913: it is clear that Henry Ford and Charles Sorensen took their inspiration from the assembly line of the Chicago slaughterhouses, with its system of winches that carried the carcasses along the path where the meat was cut up.[89] Yet the assembly line at Highland Park did constitute a breakthrough, in the sense that henceforth the process of production was entirely mechanized. Ford's "wizards of mechanics" systematized and brought to the highest point of perfection processes that until then had been separate from each other. Henceforth, it was the machine that determined the nature and the rhythm of human labor. This is why one must not underestimate the importance of Little's article and the formal introduction of unit operations in *Principles of Chemical Engineering*.

The fact is that the spread of unit operations was not a foregone conclusion, for the traditional organization of work as set up by mechanical engineers and production chemists might well have continued, as it did in Europe. The American case was certainly in stark contrast with the chemical industries of the Old World. For a variety of reasons, neither France nor Germany, nor, to a lesser extent, Great Britain, truly adopted unit operations before the middle of the twentieth century.[90]

England's chemical industry was essentially concerned with heavy chemistry, whose growth had been stimulated by the first industrial revolution. For the most part, it employed technicians and mechanical engineers, and very few chemists. Old manufacturing processes (such as the Leblanc process in the manufacture of soda), were widely used in making well-known products. As for relations between industry and the university, they were practically nonexistent, in keeping with the tradition of the independent British scholar exemplified by Charles Babbage, a prestigious figure in nineteenth-century English physics. William Perkin, the inventor of the first artificial dye (Perkin's mauve), sold his business after he had made enough money to finance his own research.[91] As Fred Aftalion summarizes it: "The leaders of the English chemical industry had lost their predecessors' dynamism and fought a rear-guard action in outdated activities."[92]

The English chemical industry did reorganize after World War I. In 1926, the creation of Imperial Chemical Industries (ICI), a cartel of some of the principal British firms, favored the development of continuous-flow chemical processes, particularly in the extremely costly high-pressure chemistry used in manufacturing ammonia, synthetic gasoline, and methanol in its plants at Billingham in northern England. At that point, the British, rather than adopting the German model, turned toward the Americans and their unit operations, thereby moving away from the traditional practices of English chemical engineering.[93] Their professional association, the Institution of Chemical Engineers, created in 1922, maintained close ties with the AIChE, and courses in chemical engineering based on unit operations were beginning to be taught at Imperial College, London, and several other universities. Unit operations were coming into their own in the chemical and petrochemical industry. Of all the major countries engaged in industrial chemistry, Great Britain was certainly the one that most rapidly adopted the new methods of American chemical engineering.

In France, the primacy of mathematics and theory in the programs of preparatory classes and the *grandes écoles* made chemistry the poor relation in the teaching of applied sciences, since for a long time it seemed resistant to mathematical model-building.[94] Eventually, in the last years of the nineteenth century, the initiative for creating the first courses of applied chemistry came from the faculties of science.[95] In this manner, institutes of chemistry directly attached to the university were created in such places as Lyon and Paris, while others were more autonomous, such as that at Nancy. In a parallel development, local concessions were made to respond to the needs of mid-sized companies; this is how institutes of chemistry came into being at Mulhouse and Rouen. Throughout the 1920s, courses in industrial chemistry remained very descriptive, as can be seen in Paul Pascal's textbook *Synthèses et catalyses indus-*

trielles (1925), based on the courses the author had taught at the University of Lille.[96]

When *Principles of Chemical Engineering* appeared in French in 1933 (under the significant title *Principes de chimie industrielle*), the French were most interested in the mathematical aspect of unit operations, for this was the feature that allowed for a better integration of chemistry into the programs of the preparatory classes and the *grandes écoles*. But since the status of chemical engineers was considerably lower than that of the graduate engineers of the École polytechnique or the École centrale who were running the French chemical companies, the chemical engineers had little influence on the organizational choices of the *polytechniciens* and the *centraliens*, even though a trade union for chemical engineers, the Syndicat des ingénieurs chimistes de France, was created in 1919. The French chemical industry, whose most dynamic elements looked to Germany rather than the United States, was incapable of conveying its precise needs to the engineering schools and universities.[97]

It was only after World War II that substantial programs based on unit operations were created, notably in Paris (the Institut du pétrole), Toulouse (the Institut du génie chimique at the University of Toulouse), and Nancy (the École nationale supérieure des industries chimiques, the former Institut chimique). The French terms for "chemical engineering" and "chemical engineer"—*génie chimique* and *ingénieur du génie chimique*—were taken over from Canadian usage. In keeping with a general trend in French industry, which just then was sending its "productivity commissions" across the Atlantic, the French chemical industry now began to model its methods of production more or less closely on those of its American competitor.[98] In Germany, meanwhile, a strict division of tasks between chemists and mechanical engineers did not favor the concept of the chemical engineer as it was being defined in the United States.[99] The German chemical industry was being built around areas that required advanced scientific efforts and yielded products with high added value, such as synthetic dyes and pharmaceuticals.[100] The chemists, who were grouped into research teams carrying out *wissenschaftliche Massenarbeit* (scientific mass production), wielded major influence from the beginning of the German chemical industry, and their knowledge, in conjunction with their authority within their firms, impressed their foreign colleagues. The French engineer Victor Cambon, who visited the BASF plants on the eve World War I, admired and worried about the "frightening" achievements of German chemistry.[101] The talent and experience of the German chemists allowed many of them to bring to fruition industrial projects of great scope. There is no question that in the 1920s, a team of chemists at IG Farben formulated precise equations for the transfer of heat and mass that closely matched the "transfer phenomena" on which the Americans were beginning to work (see Chapter 2). But this work had no organizational consequences. The very term *chemical engineer* was rejected by the Germans as hybrid and unclear.[102]

The strict division of tasks between chemists and engineers continued until the 1960s, when the first courses in chemical engineering were offered at German universities and technical institutes. This was also the time when the two major professional associations, the DECHEMA (DEutsche Gesellschaft für CHEMisches Apparatewesen) and the VDI (Verein Deutscher Ingenieure) made the promotion of chemical engineering a priority.[103] It stands to reason that at that point, the classic institutional arrangement revealed its limitations, given the accelerated diversification in the German chemical industry, which required the earliest possible adoption of new processes and the precise evaluation of production costs. Under these circumstances, the chemical engineer appeared to offer a desirable compromise, whereas in a parallel development, the star of the laboratory chemist began to fade.

Thus the European chemical industry—unlike the metallurgy industry, which rapidly, and sometimes enthusiastically, adopted Fordist methods—remained rather circumspect about unit operations until relatively late in the century. At the very time when some of Europe's most enterprising automobile industrialists, such as André Citroën, were working hard to imitate Ford's brilliant achievements, their colleagues in chemistry were looking toward the giant of the worldwide chemical industry, IG Farben, born in 1925 out of the merger of Bayer, Hoechst, BASF, and five other firms of lesser importance. Germany, not the United States, was the Eldorado of the chemical industry.

What were the peculiar features of the American chemical industry that required so many specialists trained to work in unit operations? One can distinguish two major kinds of factors: some that related to this particular chemical industry and other, more general ones that had to do with the rise in power of big American industry and with the scientific organization of work.

To begin with, the adoption of unit operations in the United States was favored by the American chemical industry's long-standing expertise in the continuous-flow manufacture of heavy inorganic chemicals. Engaging in mass production, it was in this respect ahead of its European competitors, who even before the war sometimes made use of American techniques. In 1912, for example, Americans from the General Chemical Company were hired by BASF and Bayer to build the production units for sulfuric acid at Ludwigshafen and Elberfeld in Germany.[104] To be sure, American organic chemistry was experiencing rapid growth at the end of the century's first decade, and DuPont needed new engineers to meet its needs. But Chief Engineer Everett Ackart frankly admitted that in the early 1920s, DuPont was interested above all in keeping close track of the innovations in German products and in finding clever ways of copying them and producing more at lower cost: "They sent people over to Germany—engineers went over to study and brought back all the information."[105] The Germans, for their part, accused the Americans of industrial espionage, exactly as, a few decades later, the Americans were to suspect the Japanese. Organic chemistry was characterized by batch production in situations where unit operations were less important. Because of their highly physical approach to the problems of production, unit operations were better suited to inorganic than to organic chemistry.

These characteristics made for a clear contrast between the chemical and the electrical industries. The latter had come into its own in the 1880s by bringing scientific inventions to bear on technical innovations and on the creation of powerful companies. America's major technological innovations in the prewar period essentially took place in the areas of telecommunications and electricity, to which one might add the Wright brothers' internal-combustion airplane and Maxim's machine gun. More inventive with respect to processes than to products, American chemistry may have lacked its Edison or its Graham Bell, or even its Fritz Haber, the German chemist who invented synthetic ammonia, but it nonetheless succeeded in clearly formulating its needs and inventing a new profession.

Secondly, unit operations offered a space to be negotiated between universities and industry, and by and large they responded well to the needs of both parties. Unit operations were born out of an organizational context specific to the United States, namely, the close connection between universities and businesses. As a consulting engineer and advisor to the Department of Chemistry at MIT, Little himself was in a pivotal position between the industrial and academic worlds. Walker and Lewis too were both professors at MIT and consultants for chemical firms. This interpenetration favored the emergence of unit operations, which allowed industrialists to organize their production units more efficiently and departments of chemistry or chemical engineering to organize their courses in a systematic manner. In a sense, the unit operations were a way to learn the rudiments of chemical engineering. More than electrical engineering, chemical engineering became the archetype of the interpenetration between the academic and the industrial worlds. The blurring of the frontiers between the engineer and the chemist, then, was the fruit of an arbitration and of a process of negotiation between organizations engaged in production and in the pursuit of knowledge.

Does this mean that the unit operations were invented by Little at the request of the American chemical industry? It does, in the sense that Little was a consulting engineer, and that in this capacity, he responded to the needs of industrialists whom he knew well. On the other hand, it is not easy to know whether his formulation came out of discussions with the engineers of DuPont and other firms: the archives are silent on this point. However, one can reasonably assume that Little's propositions were tried out ahead of time on a certain number of engineers, and that his report of 1915 responded to a request that industry had made to certain university departments. This was certainly the case with the petrochemical industry, which at the time was facing an explosion of demand for automobile fuel. The continuous cracking process developed by the engineer Carbon Dubbs, a graduate of MIT, which replaced batch processing in 1924, was based on the principle of unit operations. Petroleum producers were the first to benefit from Little's propositions. The new methodological tool contributed to giving a solid grounding to the profession of chemical engineer, which henceforth had great potential in a rapidly expanding industry. In 1922, Arthur Little presented a report to the AIChE in which he recommended that academic programs be organized in terms of unit operations. The association reacted very favorably—one of its officials called the report "monumental"—and proceeded to draw up a program designed to encourage the establishment of standardized teaching.[106]

DuPont's directors too were quick to understand the importance of the new concept and encouraged its propagation by the engineering schools. Charles Reese was frequently invited to speak to engineering students who might some day work for DuPont, and in the 1920s, he never failed to refer to unit operations and to insist on their operational value. On 5 May 1922, at the University of Virginia, for instance, Reese evoked the need to attack problems from a "theoretical" standpoint rather than relying on "experience." He also stressed the role DuPont had played in the adoption of unit operations, which "require scientific training of the highest order, such as the work being done by Professor Warren K. Lewis of MIT."[107] Reese continued to promote unit operations when he became president of the AIChE in 1923. Moreover, DuPont set up working groups in its principal plants for the purpose of studying the adoption of unit operations. This did not necessarily mean doing away with the logbooks, which were very useful for transmitting experiences; rather, it was a matter of codifying practices and obtaining more scientific data than in the past. For example, instead of relying on an operator, a thermometer, and an emergency spigot for keeping control of a batch of nitrocellulose, a working group tried to understand why the temperature sometimes became too high.

The concept of the unit operation became important for DuPont to the extent that it facilitated its transformation from a gunpowder manufacturer into a diversified company, particularly in the area of cellulose-based products, but also, as we shall see, in a sphere where the chemical engineers were to rule, that is, in high-pressure chemistry. Nonetheless, until about 1920, DuPont—along with the rest of the American chemical industry—was not yet capable of standing up to its German competitors. The turning point was to come later, when DuPont became engaged in the development of an entirely new product, nylon, a venture in which research and development yielded innovations in both the process and the product.

Before this came to pass, the trial by fire for American chemical engineers was the manufacture of poison gas during World War I, specifically from 1917 on. This was an opportunity for the MIT professors and their students to put their knowledge into practice and to give more visibility to their profession. They half-succeeded.

The Manufacture of Poison Gas During World War I

The production of lethal gas in the United States mobilized several thousand chemists and chemical engineers. Even though the United States manufactured relatively little gas by comparison with the European belligerents (7,200 tons, as against almost 100,000 tons for Germany, and 31,000 tons each for Great Britain and France), the American government created huge facilities to do so, which began to produce massive amounts during the last months of the war. Above all, this was an opportunity for the chemical engineers, particularly those at MIT, who had been mobilized for this effort, to show off their know-how to the country's scientific and military establishment. It is regrettable that the production of these gases has still not been made the object of historical studies, as if the only aspects that mattered were the laboratory phase and the military use of poison gas.[108]

On 22 April 1915, on the Ypres front in Belgium, a German unit especially assembled for this occasion, opened some 6,000 cylinders of chlorine under the watchful eyes of officers of the General Staff and the chemist Fritz Haber, who had developed the procedure. On that late afternoon, a cloud of yellowish gas, slowly pushed along by the breeze, soon reached the French positions and suffocated the soldiers who had not fled in time. The image of these unfortunate men, taken by surprise and asphyxiated, was to leave an indelible mark on public opinion everywhere and made poison gases the symbol of a horrible war from which all ethical principles had been banished.[109]

Thereafter, the troops were quickly outfitted with rudimentary gas masks, and the Allies too launched a program of manufacturing poison gas. Both sides mobilized their best chemists, who tried to outdo each other in the race to develop organic chemical gases that would be more effective than chlorine. Among them were phosgene and especially the infamous mustard gas, which was introduced by the Germans in July 1917 and killed by contact rather than only by inhalation, so that gas masks became partially ineffective. A French invention, soon imitated by everyone, consisted of packing the gas into shells specially designed for this purpose. By late 1917 and in 1918, poison gases were used on a massive scale by all the belligerents.

The Americans, though initially skeptical about the effectiveness of poison gases, soon began to gear up for production. In early 1917, the Bureau of Mines took the initiative to conduct preliminary research, seconded by the American Chemical Society and the chemistry section of the National Research Council (NRC). Toward the end of the conflict, this program was placed under the responsibility of the Army through the creation of the Chemical Warfare Service (CWS). The Bureau of Mines was charged with manufacturing gas masks for U.S. troops, while the NRC organized research on the gases at twenty-five university laboratories. But time was short, and in June 1917, the NRC decided to build a huge complex of laboratories in Washington, D.C., on the campus of American University, a recently founded institution that still had vacant space. When the Armistice came, almost 1,700 chemists were working in this research center, called the "American University Experiment Station." To be found there were many chemical engineers and chemists, among them Warren K. Lewis of MIT, James Conant of Harvard (a future president of that university),[110] and Roger Adams of the University of Illinois. Altogether, 4,003 of the 5,404 American chemists who were called to the colors were employed in connection with this chemical warfare program.[111]

In addition to the poison gases already used by the Europeans—such as hydrogen cyanide (the famous *vincennite* of the French), phosgene, chloropicrin, and mustard gas—two new types were manufactured by the Americans, adamsite, developed by Adams and made from chlorine and arsenic, and lewisite, invented by Winford Lee Lewis, a chemist at Northwestern University (not to be confused with Warren K. Lewis).

But it was not enough to concoct these deadly mixtures; they also had to be manufactured in large quantities. No company was interested in this job, for all of them were already very busy filling orders for the war effort, not to mention the fact that manufacturing highly toxic products was dangerous. As we shall see, arguments of this kind were also advanced by DuPont a quarter century later, when it was asked to produce plutonium. In 1917, the U.S. government therefore took over the manufacturing of gases itself and began by building a large arsenal near Baltimore, Maryland, in November of that year. Edgewood Arsenal, as it was called, comprised 558 buildings and employed more than a thousand workers to produce the gases and fill shells with them.

The men responsible for production at Edgewood Arsenal were chemical engineers. William Walker, holding the rank of colonel, was appointed to head the arsenal on 1 March 1918, while Warren K. Lewis, who remained a civilian, directed one of the research and development divisions of the laboratory at American University. William McAdams, one of the future authors of *Principles of Chemical Engineering*, at the time still a graduate student at MIT, was made a captain and charged with producing mustard gas in an experimental plant at Cleveland and then at Edgewood, in collaboration with James Conant. The undergraduate MIT students hired by Walker were given the rank of lieutenant. These engineers were placed under the authority of Arthur Noyes, professor of chemistry at MIT and president of the NRC. The program of gas production mobilized the most advanced engineers, the crusaders for chemical engineering—men who, aside from patriotic motivations, saw this venture as an opportunity to demonstrate their competence.

As head of the arsenal, Walker was free to organize it as he saw fit. At the industrial site, the plants for making chlorine, phosgene, chloropicrine, and mustard gas were grouped together, along with the units where 75 and 55 mm shells and grenades were filled.[112] A report of March 1919 provides a few details of the phases in the arsenal's construction. The chemical aspects of the manufacture of the gases were not complex (phosgene, for example, is a mixture of chlorine and carbon monoxide aided by a catalyst), but the toxicity of the products made the operations extremely dangerous. To be sure, the filling operations were carried out automatically by means of compressors that filled the projectiles, which were arranged in rows of five on a moving cart—unlike at the French factory in Aubervilliers, where the filling was done by hand—but several hundred workers were poisoned, some of them fatally. The working conditions were very harsh: the men labored in burning, acrid smoke, barely protected by rudimentary masks. And when the Spanish influenza epidemic struck, their already weakened systems were unable to withstand it, and the virus decimated the Edgewood Arsenal.[113] Under these circumstances, it was not easy to recruit workers, particularly since there was stiff competition from private firms in the area. Despite premiums and high wages, turnover exceeded 30 percent per week, and accidents became even more frequent. Direct testimony is lacking, but one can imagine the experiences of the students and professors of chemical engineering who worked at Edgewood. One of them, Dana Demorest, who may have been more cynical than his peers, subsequently noted in an article in the review *Chemical Warfare* that this had been a great learning experience.[114] Much of the equipment had had to be designed on the spot, he explained, since certain products, such as pure carbon monoxide, had never been manufactured in industrial quantities. Demorest furnished several examples of practical problems in chemical engineering that the engineers had tackled.

They worked ceaselessly to enable the United States to catch up with the other belligerents. Walker was proud to have further improved an originally French method of mustard gas production that was much more efficient than that of the Germans. By November 1918, Edgewood produced more gas in a day than Great Britain and France did in a month. "If the Armistice had not cut our momentum," Walker sighed, "we would have produced 100 tons a day, even 200, by spring 1919." The same report of 1919 also states that "it is to be regretted that this magnificent unit did not produce at its maximum rate, as planned."[115]

The industrial production of chloropicrin began in June 1918, that of phosgene in July, and that of mustard gas in September of that year.[116] As for lewisite, which was potentially more effective than any other gas, a first shipment of it left Baltimore in early November 1918. The Armistice surprised the vessel in mid-ocean, and the lewisite was thrown overboard.[117] Altogether, the Americans manufactured about 1,800 tons of chloropicrine, 1,400 tons of phosgene, and 900 tons of mustard gas. Edgewood was barely beginning to produce at full capacity when hostilities ended.

But the arsenal was not dismantled: later on, it produced tear gas used by police forces all over the world, and even lethal gas for judicial executions (the first execution by gas took place in Nevada in 1924). Nor was the production capacity for military gases abandoned, for at the insistence of certain military men, chemists, and chemical engineers like Walker and Lewis, the United States did not sign the 1925 Geneva gas protocol outlawing the use of poison gases. The CWS continued its activities, supported by the association of former employees at Edgewood and Washington (the Chemical Warfare Reserve Officers Association),[118] as well as by the American Chemical Society, which officially condemned the Geneva protocol. A monthly publication, *Chemical Warfare*, published by Edgewood, informed its readers about the activities of the arsenal and the promotions of reserve officers, and also published propaganda articles on the desirability of using poison gas, which it touted as more humane than ordinary artillery, and in any case necessary, particularly in combating "half-civilized enemies."[119] In addition, the review insisted on the civilian application of the gases: "phosgene, the most lethal of the lethal gases, is becoming increasingly important in the manufacturing of the most brilliant dyes, pink, purple, green, and yellow."[120] Military and economic interests combined their efforts to warn of the German danger and enlist the support of its readers, who may be assumed to have been already on the side of the CWS. Published between 1919 and 1934, and then replaced by a trimestrial bulletin that vegetated until 1945, *Chemical Warfare* was unable to convince the majority of Americans, either civilian or military, of the interest or the nobility of its cause.

Nonetheless, the CWS can been analyzed as the first significant example of large-scale collaboration between the military and American scientists. The Armistice of 1918 did not put an end to their partnership. The informal "sleeper" network that had been put into place would be reactivated by World War II. It was no coincidence that in 1941, James Conant was appointed to head the National Defense Research Committee (NDRC), the government agency that oversaw the Manhattan Project, and that a certain number of former Edgewood scientists returned to government service during World War II. The mobilization of scientists in the Manhattan Project must therefore not be seen as a radically new undertaking, as certain historians too often claim.

In the 1920s, a program for the training of CWS officers was set up in the department of chemical engineering at MIT in cooperation with Edgewood's Chemical Warfare School.[121] A two-year program, it called for a series of courses in chemical engineering, mathematics, organic and inorganic chemistry, and industrial organization, with special emphasis on learning about unit operations. References to the "return to normalcy" that allegedly followed the Great War must certainly not be taken literally, as if the war had been nothing more than a quickly closed parenthesis. America might have gone back to normalcy, but normalcy had changed.

This being the case, one understands the very rapid and sometimes spontaneous mobilization of chemists from the very beginning of World War II. To cite only two examples, in 1941, John Woodhouse, a young chemist in DuPont's Ammonia Department, together with a few of his colleagues, developed a flammable liquid for use in fire bombs and flame throwers, after he had spontaneously contacted the directors of Edgewood Arsenal.[122] And the chemical engineer Robert Pigford, one of Woodhouse's colleagues at DuPont, was in charge of manufacturing an improved mustard gas that in the end was never used.[123] It is therefore not surprising that for a long time DuPont was reputed to be a manufacturer of poison gas. According to *Time* magazine, the Japanese, believing that the Americans were preparing to use gas against them, "answered saucily. They said that if the U.S. uses gas, Uncle Sam's boys will be given a smell of their own Du Pont gas which the Japanese captured at Guam."[124]

Both in the United States and in Europe, World War I had become "the chemists' war." During the war years and immediately afterward, the chemical engineering students at MIT referred to their work at Edgewood Arsenal as a milestone in their careers. But the arsenal's operations probably did not live up to their expectations, in part because the war ended just as Edgewood began to produce in significant quantities, but above all because the hateful image associated with poison gases made it very difficult to use this as the foundation of a positive attitude toward chemistry and chemical engineering. Unlike in the post-Hiroshima era—when physicists and to a lesser degree chemical engineers enjoyed an extraordinary, albeit ambiguous, prestige—the chemists and chemical engineers of the Great War willy-nilly had to keep a low profile. There was just no way to promote chemical engineering against the background of mustard gas in the trenches. This was clear to the engineer John Hammond who, a few years later, in a small book for the use of students, deplored the fact that "mentioning chemical engineering and the war immediately makes people think of gas, of shells, and of poison gas bombs. These seem to be the chemical fruits of the World War."[125] But then Hammond went on to explain that the chemical engineer did work on other things, and indeed that his profession was necessary to most of the technological activities of the day. In short, the crusaders for the young profession felt that they had to promote chemical engineering by referring to civilian consumer goods, even if the manufacture of war gases had served as a practical apprenticeship for a whole generation of engineers.

The Creation of a Discipline: Chemical Engineering at MIT

After it had reorganized its structure in the early years of the century in order to adapt to the changing scale of the American economy, MIT played a crucial role in the rise to power of American chemical engineering. Its training in this field and its ties to DuPont became the paradigm of a way of acquiring technical knowledge that was instituted—not always without opposition—for the purpose of producing the engineers that DuPont and other firms felt they needed if they were to make the American chemical industry the equal of its powerful German competitor and possibly to surpass it some day. DuPont, which relied on MIT for recruiting the elite of its chemical engineers, closely watched the school's activities.

But the relations between the business world and the major research universities were not always smooth.[126] One can easily overestimate the harmony in relations between the universities and big business by being too preoccupied with demonstrating their collusion in political and social matters.[127] Conflicts did arise, both among academics and between academics and the directors of major companies. In this respect, the case of chemical engineering appears to be similar to that of electrical engineering. The cooperation between business and the academy was not a matter of course, but it was eventually and reluctantly achieved on the basis of compromises elaborated in the 1920s.

At the time, MIT was in a transitional period in its history, for it was no longer the regional technical institute of its beginnings and not yet the major scientific and technical university that was to emerge a few years later as a pillar of the country's military, scientific, and industrial complex.

In the United States, the advance of new engineering specialties was so rapid that by the end of the 1920s, there were 50 percent more students in electrical than in mechanical engineering. As for chemical engineering, its hesitant beginnings notwithstanding, it attracted at that time the equivalent of half

Table 1.1   Chemical engineering students in the United States, 1910–1960

Year	Chemical engineering students	Chemical engineering students as % of all engineering students
1910	869	3.7
1920	5,743	?
1936	12,550	20
1946	17,392	14
1950	13,647	8.4
1960	12,704	9.1

Source: *Chemical Engineering Progress,* various issues.

the students in mechanical engineering. Let us consider the advances over a half century. Table 1.1 shows the number of students in chemical engineering between 1910 and 1960 and their percentage in relation to the total number of engineering students at all American colleges and universities.[128]

Note that the number reached 20 percent in 1936 and then declined after World War II, no doubt because of the rise of new specialties (aeronautics and electronics), which became the beneficiaries of the financial manna distributed by the U.S. Department of Defense and its affiliated agencies, and which attracted many enthusiastic young people. Nonetheless, the advance of chemical engineering is remarkable, considering the strides made by the engineering profession as a whole. One academic institution, MIT, played a decisive role in its rise to power. If we are to understand the evolution of chemical engineering at MIT, we must place it into the context of the general history of MIT and its internal rivalries.[129]

MIT had been founded in 1861 by a group of enlightened reformers, who were not only engaged in the abolitionist movement but also wanted to promote "the practical sciences," in contrast to the traditional universities, which by and large neglected technical education.[130] Following an early period marked by certain financial difficulties, MIT rose to eminence in the 1880s and 1890s under the presidency of Francis Walker, thanks to a new group of dynamic professors who expanded its curriculum and established ties to local industry. Even at this early stage, experienced engineers gave courses, students had access to training periods in industry, professors served as consultants in various places, and new training programs were launched (as part of this innovative trend, MIT had created its first courses in chemical engineering in 1888).

But by the beginning of the twentieth century, MIT was compelled to re-evaluate its highly pragmatic, local approach by financial difficulties, the rise of research universities such as Johns Hopkins and the University of Chicago, and the creation of industrial research laboratories. It was time to redefine its mission: should it become a college for undergraduate engineering students or a research university? At this point, a struggle between several factions in the faculty ensued.[131] Three different coalitions of professors and administrators were involved. The first was committed to MIT's traditional mission and saw it as a regional engineering school, but this view was already outdated at the beginning of the century and rapidly lost ground thereafter.

The second, which was grouped around Arthur Noyes, professor of physical chemistry, was bent on turning MIT into a major scientific university, with a focus on the hard sciences. Noyes believed that the best way to train engineers for constantly evolving fields was to begin by giving them the most thorough scientific grounding possible. This might initially make it more difficult for freshly minted MIT engineers to adapt to the industrial world, but Noyes felt that their scientific training would benefit them in the long run. Hence the Research Laboratory in Physical Chemistry, or RLPC, created by Noyes in 1906, was explicitly designed to train elite engineers who could call on solid knowledge of physics and chemistry. Significantly, the RLPC awarded MIT's first Ph.D. in 1907.[132] Noyes and his friends were also in favor of establishing close ties with the laboratories of major companies, such as DuPont, General Electric, and AT&T.

The third faction, consisting mainly of professors of electrical and chemical engineering, among them Dugald Jackson in electrical engineering and William Walker, the former associate of Arthur Little and head of the Edgewood Arsenal, saw the engineering profession in global terms, along Taylorist lines: Walker pleaded for engineers to involve themselves in the country's social affairs and prepare to become managers. In his opinion, MIT should be not so much a university as an engineering school, and its primary missions should be to train leaders for tomorrow's industrial world and to contribute to needed social reforms. But Walker was willing to subordinate teaching to the exigencies of the industrial world, rather like the future president, the engineer Herbert Hoover, and quite unlike Progressive engineers such as Morris Cooke, a follower of Taylor, who felt that the engineers had a social mission that was fundamentally opposed to the interests of big capital.[133]

In opposition to Noyes, Walker wanted to anchor teaching in practical rather than theoretical problems. Not, of course, that Walker was hostile to scientific research, as the historian John Servos has pointed out. Being himself a graduate of Göttingen University in Germany, he was aware of the importance of thorough training in chemistry. But Walker thought that the students should come into contact with industrial problems as early as possible, understand the constraints of production and financing with which companies had to deal, and be fully operational as soon as they had their diplomas, so that they could rapidly become integrated into the network of American industries. This issue of how to train engineers has been raised again and again in the United States and elsewhere. Its implications are not only technical but also social and political, for it was important to make sure that future upper-level employees in industry would not defect, and that their training guaranteed both their competence and their loyalty to the system.

In order to win acceptance for his position against Noyes in the Department of Chemistry and Chemical Engineering, Walker proceeded in several stages. First, in 1908, he created the Research Laboratory of Applied Chemistry, or RLAC, a semi-autonomous laboratory within the Department of Chemistry, which received the bulk of its funding from industrial sources. Walker's objective was to promote chemical engineering and to develop a doctoral program in this field. The laboratory was almost immediately successful, and in 1909, twice as many students registered in chemical engineering as in chemistry. In 1920, the department's brochure stated with satisfaction that "at no time within the history of our country has public attention been so focused upon the development of our chemical industry as it is today. Never has the demand been so insistent for chemical engineers competent to develop and direct these industries."[134]

In 1919, the conflict between Walker and Noyes—compounded by personal and social considerations—led to the departure of the latter. Noyes joined the Throop Polytechnic Institute in Pasadena, California, soon to become famous as the California Institute of Technology thanks to the leadership of the astrophysicist George Ellery Hale, Noyes, Robert Millikan, and Michael Pupin, the last two having defected from the University of Chicago and from Columbia University, respectively.[135] Noyes's departure was a heavy blow to MIT's Department of Chemistry, and indeed to its scientific activities in the narrow sense. The best young chemists no longer chose a university that had become almost exclusively turned toward large and often small industry, and whose subcontracting activities sometimes seemed to take precedence over its primary mission as a training institution. The institutional consequence of the break between chemists and chemical engineers was the creation of a Department of Chemical Engineering in July 1920. This was the first autonomous department of chemical engineering in the United States and a triumph for Walker and his colleagues in chemical engineering, who were still basking in the glory of their work at Edgewood Arsenal.

Richard McLaurin, president of MIT from 1909 to 1920, was on very good terms with Walker, but he wished to establish closer ties with the major companies, instead of continuing to rely on local industries. The du Pont cousins, who were MIT alumni, were asked to contribute to MIT in 1911, as was George Eastman, the founder and director of Kodak, who turned out to be particularly generous (he was to give more than $25 million to MIT between 1912 and 1925).[136] One DuPont engineer put it succinctly when he said: "You see, there was a leaning toward MIT men because you had Pierre, Irénée, Lammot, and Coleman."[137]

MIT's Technology Plan, instituted in 1919 and administered by the Division of Industrial Cooperation and Research, headed by Walker, was designed to create even closer ties with industry by getting businesses to pay a lump sum for technical help and access to the MIT library.[138] The aim was to institutionalize and formalize relations with the industrial community, and also to refill MIT's depleted coffers. There is no question but that the Technology Plan was the logical conclusion to several decades of MIT's collaboration with companies, local at first and then national, and it quickly recruited 200 firms, some of them, such as General Electric, AT&T, U.S. Steel, and U.S. Rubber, representing big business.[139] It is noteworthy that DuPont did not sign up.

The Curriculum

In the early 1920s, chemical engineering was a rapidly growing specialty. Table 1.2 shows the number of engineering students inscribed at MIT between 1907 and 1920 by specialty. Chemical engineering appears in second place for the academic year 1919–20, right behind mechanical engineering. Even though this fact is only relative, since in the course of the 1920s, students in electrical engineering were to outnumber those in chemical engineering, it remains true that chemical engineering was the field that showed the most spectacular growth in the first two decades of the twentieth century. By 1920, chemical engineering had become popular. From the century's first decade on, and as

Table 1.2   Undergraduate students in engineering at the Massachusetts Institute of Technology, 1907–1920

Years	Civil engineering	Mechanical engineering	Mining engineering	Electrical engineering	Chemical engineering	Total
1907–8	210	227	118	202	59	908
1908–9	197	197	104	209	71	884
1909–10	207	204	99	203	84	926
1910–11	220	198	90	210	128	953
1911–12	217	214	79	203	129	961
1912–13	212	243	50	201	149	987
1913–14	209	279	37	196	141	1,003
1914–15	197	271	34	205	146	1,057
1915–16	188	279	46	235	157	1,165
1916–17	172	270	55	233	173	1,179
1917–18	160	210	40	186	164	983
1918–19	111	172	40	135	155	867

Source: MIT, "Annual Report of the President," January 1921.

a logical consequence of the new preeminence of chemical engineering over chemistry, most students entered MIT with the intention of becoming chemical engineers rather than chemists, and this trend only grew during the interwar period (see table 1.3).

The training of a chemical engineer normally took four years, which corresponds to the years of college leading up to a bachelor's degree. Some of the students spent another year to obtain a master's degree, but that was as far as they could go, for until the early 1920s, it was not possible to obtain a doctorate in chemical engineering in the United States. The first such doctorate was awarded by MIT in 1923. This situation was not without parallels in the fields of chemistry and physics, for until World War I, students in chemistry generally had to go to Germany if they wanted to obtain a doctor's degree. All of the great American chemists of the early twentieth century, whether they were industrialists or academics, had obtained their doctorates in Germany. Willis Whitney, director of the research laboratory at General Electric, had worked with Willem Oswald at Leipzig University; Charles Reese, director of the Du-Pont laboratory, had obtained his doctorate at Heidelberg. This was true in physics as well; think of J. Robert Oppenheimer and Edward Condon, who in the 1920s had been students at Göttingen and Munich, respectively. Until the 1930s, the German universities were an obligatory way station for the best

Table 1.3   Bachelor's degrees awarded in chemistry and chemical engineering at the Massachusetts Institute of Technology, 1885–1934

Years	Chemistry	Chemical engineering
1885–89	38	0
1890–94	50	31
1895–99	98	49
1900–04	78	51
1905–09	82	65
1910–14	50	132
1915–19	63	187
1920–24	52	419
1925–29	81	238
1930–34	71	240

Source: MIT, "Annual Report of the President," various years.

American students in such fields as organic chemistry or quantum physics, whether they were working on a doctorate or doing postdoctoral research. In the case of chemical engineering, however, such stays in Germany were not essential.

The MIT Chemistry Department's internal struggles between Walker and Noyes suggest that the content of the course of study for chemical engineering was at the core of the debate there. The opposition between supporters of advanced scientific training and those who favored more practical training matched the opposition between small and large chemical firms, which did not need the same kinds of engineer. The organization of what became the Department of Chemical Engineering in 1920 and the conceptual definition of that new discipline reflected this opposition (see fig. 1.2).

Following the reorganization of the program of industrial chemistry in 1905 and the opening of the Research Laboratory of Applied Chemistry three years later, Walker and his young colleague Warren K. Lewis proceeded in 1916 to create the School of Chemical Engineering Practice, whose teaching programs combined academic courses and prolonged training periods in industry.[140] Even more than the RLAC, the School of Chemical Engineering Practice was focused on small chemical firms. MIT's Department of Chemical Engineering (founded in 1920), thus brought together the RLAC and the School of Chemical Engineering Practice, and had the closest ties to industry, along with Dugald Jackson's Department of Electrical Engineering (Jackson had formed a

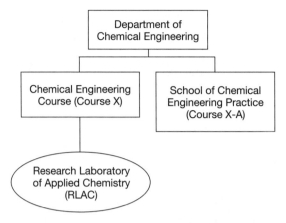

Fig. 1.2. Chemical engineering at the Massachusetts Institute of Technology in the early twentieth century

partnership with General Electric in 1917).[141] In this sense, chemical and electrical engineering were the technical specialties that most clearly embodied the interpenetration between the industrial and the academic worlds.

The School of Chemical Engineering Practice divided itself into three divisions, each located at an industrial production site. One was a textile division in Bangor, Maine (the Eastern Manufacturing Company and the Penobscot Chemical Fibre Company), another was situated at Buffalo, New York, and associated with a small steel company and a detergent factory (the Lackawanna Steel Company and the Larkin Soap Company); and the third was at Everett, Massachusetts, and functioned in collaboration with the Merrimack Chemical Company, the Revere Sugar Refinery, and the Boston Rubber Shoe Company (which respectively produced dyes, sugar, and rubber). This linkage between the School of Chemical Engineering Practice and small firms was in keeping with Walker's wish to contribute to the development of a dense and efficient industrial network in New England, which, in his opinion, required that small firms also have access to the latest research and to the means of improving their competitive edge.

The engineers trained at the School of Chemical Engineering Practice had a very practical approach to problems: starting out as production engineers, they went to work as soon as they had their diplomas and were highly sought after by small companies. All of MIT's engineering students were trained in the unit operations methods developed by Walker and his colleagues in *Principles of Chemical Engineering.* Unit operations were carried out in the most practical manner, either by following "recipes" or, conversely, by applying underlying theories of mathematics or physics.

DuPont was prominent among the companies that contributed to the solid positioning of chemical engineering at MIT and other universities but was not equally interested in all the subdivisions of MIT's Departments of Chemistry and Chemical Engineering. If the School of Chemical Engineering Practice, designed to teach undergraduates, was mainly of interest to smaller and relatively low-technology companies, the Research Laboratory of Applied Chemistry was in a different category, having engaged in true research since 1920 under the direction of Thomas Haslam and directed its teaching to graduate students and the best undergraduates. MIT's archives show that DuPont privileged the research of graduate students rather than activities at a lower level. In other words, DuPont was more interested in the RLAC than in the School of Chemical Engineering Practice, whose practical training in maintenance and repair of equipment was designed for smaller companies.

MIT's courses in chemical engineering soon came to be considered the standard to which all other teaching institutions should aspire, as a report commissioned by the AIChE and published in 1922 stated.[142] Written by a committee chaired by Arthur Little and consisting of five other members, including Charles Reese, the report presented the findings of a survey of seventy-eight institutions "known to offer courses in chemical engineering." Seventy-three of them had responded. The report expressed alarm that half of the 210 courses offered were "neither necessary nor desirable." It recommended following the example of ten universities, which offered between 35 (Michigan) and 62 (Columbia) courses and commended MIT and its School of Chemical Engineering Practice in particular for "the eminent results already achieved and the exceptional enthusiasm of the students." Appended to the report was a description of the activities of this MIT division. Finally, the committee presented a definition of chemical engineering that tied it very closely to unit operations, which was a way to urge all departments of chemical engineering to revamp their courses accordingly. Thus it appears that the reduction in the number of courses that the committee recommended was related in most cases to its intent to generalize the use of unit operations, in other words, to organize a body of knowledge into synthetic units rather than lay it out in a series of descriptions.

DuPont and MIT

In forging close ties to MIT and a few other universities, DuPont's essential objective was to recruit chemical engineers whose training would meet its needs. DuPont's relations with MIT materialized in a threefold manner: through financial support, the submission of research topics, and the hiring of consultants. By way of financial support, DuPont established a fellowship program for professors and engineering students. Averaging $500, these grants were distributed among departments of chemistry and chemical engineering at selected universities. Until about 1927, DuPont provided several fellowships to each of these departments, which were then free to distribute them as they saw fit. Eventually, DuPont set up a fellowship committee in order to target these grants more precisely. Between 1918, when the first fellowship was set up, and 1932, 329 fellowships were granted, for an annual total cost of $22,500. By 1952, 1,037 fellowships were awarded annually, at a cost of $275,000.[143]

MIT enjoyed special treatment, first, because its Department of Chemical Engineering was undoubtedly the best in the country; secondly, because the du Pont family had personal ties to it; and, thirdly, because it was in financial difficulties. In 1930, under the presidency of Karl Compton, MIT approached captains of industry—particularly among its alumni—for contributions. Lammot Jr. and Pierre S. du Pont agreed to contribute $25,000 each for ten years to MIT's Technology Loan Fund, while Irénée and T. Coleman du Pont gave $25,000 for the year 1930 only. On the list of generous donors, one also notes the names of Alfred Sloan, president of General Motors and a close friend of the du Ponts ($50,000 in 1930), Gerard Swope, president of General Electric ($25,000 per year for ten years), and George Eastman, founding president of Kodak and main contributor to MIT ($50,000 annually for ten years, to which must be added many previous donations, including a gift of several million dollars Eastman gave anonymously on the occasion of MIT's move to Cambridge in 1916).[144] In point of fact, the bulk of the Department of Chemical Engineering's budget came from private sources. In 1921, for instance, only $11,000 of the $70,000 budget of the RLAC was provided by MIT itself.[145]

In return for these outlays, DuPont expected several things, foremost among them the exploration by the universities of a certain number of topics that were of particular interest to the company. In 1920, DuPont's Department of Chemical Research suggested some forty research topics in chemistry and chemical engineering, on which some of the recipients of fellowships started to work, although they were not, strictly speaking, obliged to do so.[146] Reading the different reports of MIT's Departments of Chemistry and Chemical Engineering is very enlightening, for one can see that some of the research topics were related to DuPont's preoccupations. The report for 1925, for instance, mentions that in 1924, research had been done on gases subjected to high pressure and on the thermodynamic properties of ammonia mixed with nitrogen. At the time, such research was a priority for DuPont, which, as we shall see, wished to launch the large-scale production of synthetic ammonia and nitric acid and had encountered a certain number of technical difficulties.[147] The MIT report specifically stated that high-pressure chemistry "had aroused a great deal of interest," which was a way of saying that the department's financial backers, particularly DuPont and the oil companies (which were interested in the cracking of hydrocarbon molecules), had pushed in that direction. The reports of MIT's Board of Trustees ("The Corporation") also provide an idea of the priorities of the leaders of industries, among them DuPont, which was represented first by Pierre and then by Lammot du Pont, who were also members of the visiting committees of the Departments of Chemistry and Chemical Engineering.

These two industrialists regularly insisted that priority must be given to research, which seems to indicate that they were not in favor of having MIT become a simple provider of services to companies that did not have the means to conduct their own research.

Lammot du Pont spoke up, for instance, at an MIT board meeting on 8 January 1930, pleading for an increase in research funds and for new buildings for the Departments of Chemistry and Chemical Engineering. In support of his demand, he promised a gift of $100,000.[148] The reports of the board indicate that DuPont definitely favored MIT's turn toward high-level research activities, rather than toward a role as provider of services for industry. Lammot du Pont expressed this in a letter of 2 January 1930, saying that "MIT should refuse to undertake research work for corporations, unless the problem is one of a fundamental nature, will be an aid to the industry as a whole as distinguished from the particular corporation proposing the work, will be of such a nature as to help the staff and students from the educational point of view."[149] Of special interest to DuPont was the mathematical formalization of unit operations. This priority seems to be confirmed by Charles Reese, director of chemical research at DuPont until 1923 and thereafter president of the AIChE, who also stressed the need for the scientific training of engineers. "Practical training is important but not essential in all cases. In my experience I have found that men thoroughly trained in the most fundamental science of chemistry after a few years of work in an industrial research laboratory make the most efficient chemical engineers" Reese told a meeting of the American Society of Mechanical Engineers at the University of Virginia in May 1922.[150]

This attitude is actually quite close to that expressed by Noyes in the 1910s, before he was marginalized by Walker and moved to Caltech. It should be noted that Reese had been a member of the Arthur D. Little committee, which, in its study of the training of chemical engineers commissioned by the AIChE, had advocated a highly practical training, like that introduced by Walker at his School of Chemical Engineering Practice. But Reese does not seem to have agreed with all the conclusions, as he hinted in the same address.

This view was to come to the fore once again after World War II, when the chairman of DuPont complained openly that MIT and some universities were now nothing but engineering consulting firms.[151] DuPont always urged universities to devote their efforts to giving engineers a good scientific grounding; the firm would later take care of acquainting them with the industrial realities.

DuPont also hoped that its fellowships would serve to identify and attract the best students in chemistry and chemical engineering. The natural sequel to working on a project of interest to DuPont with the financial support of the company was to be hired by one of the firm's departments. This was not true only of MIT. Many other universities asked DuPont to participate in joint committees composed of professors and representatives of chemical companies charged with evaluating teaching programs and, if necessary, adjusting them with a view to the needs of industry. Rutgers University in New Brunswick, New Jersey, for instance, asked DuPont for this kind of cooperation on 16 January 1940. After an exchange of memoranda between Lammot du Pont and Charles Stine, the Executive Committee's vice president for research, DuPont decided to send William Calcott, at the time director of the Jackson Laboratory and in charge of hiring for that department, which specialized in dyes. Calcott, an expert in corrosion, was known best as the author, with John Perry, of the *Chemical Engineer's Handbook,* a reference work for engineers and engineering students, to which more than sixty of DuPont's chemical engineers had contributed.[152] His presence on the Rutgers evaluation committee allowed DuPont to bring its weight to bear on the university's choices for teaching and research and in return offered Rutgers a better opportunity to place its students with DuPont.

More generally speaking, DuPont in 1928 endeavored to rationalize its hiring procedures in order to avoid unnecessary competition for the best candidates among its departments. In a report concerning this matter, the head of the Personnel Department, F. C. Evans, noted that in 1928, representatives of DuPont had visited forty-five universities, compared to thirteen in 1926 and twenty-four in 1927. Also in 1928, fifty chemists and chemical engineers had been hired as a result of these visits, compared to twenty-two in 1926 and forty in 1927.[153]

Aside from MIT, the other important university in the relational network established by DuPont was the University of Illinois, whose Department of Chemistry taught both chemistry and chemical engineering. The chemical engineers graduated by this university were therefore very well trained in chemistry, and organic chemistry in particular, which was Illinois's forte. Two renowned professors of organic chemistry there, Roger Adams and Carl "Speed" Marvel, built up their department's great reputation and established solid ties with DuPont.

In addition, Adams and Marvel were also—and this is the third aspect of the relations between the company and the universities—consultants for DuPont. They took turns going to Wilmington every month for informal meetings with the chemists and chemical engineers of the Department of Chemical Research. These special ties to DuPont also helped them place their students, for 46 of Marvel's 146 doctoral students went to work for DuPont after they obtained their degrees, as did a large number of Adams's 184 doctoral students.[154]

Both of these professors specialized in the study of the polymers, one of the fields of interest to DuPont. Marvel, who himself had become interested in polymers thanks to Wallace Carothers, the inventor of nylon, in turn steered a certain number of his students toward polymer chemistry with the financial support of DuPont. The University of Illinois, along with MIT, was the mainstay of DuPont's technical support system, the former for chemistry, the latter for chemical engineering—even if, as we shall see in Chapter 2, DuPont blurred the borders between chemistry and chemical engineering by enriching each discipline with elements of the other. In the beginning, the consultants were required to sign a contract of exclusivity with DuPont, but this was no longer the case after World War II, because by then academics were no longer willing to work exclusively for one company.

There is no question that consulting academics benefited the company, but it is not easy to measure their contributions precisely, except for the students they steered toward DuPont. John Woodhouse, an engineer in the Ammonia

Department, recognized that the consultants were "morale boosters," but that most of them did not "contribute ideas for research,"[155] except for Adams and Marvel, the very emblems of a decidedly fruitful cooperation.

DuPont's engineers stayed in touch with the academic world through the consulting professors, through the courses that some of them taught at a university, through their participation in university committees, and also through their professional publications, which carried a flood of technical information, job offers, and so on. Some of DuPont's engineers might even choose to return to the university. Alan Colburn, a chemical engineer hired by DuPont in 1929, left the company in 1938 to become head of the Chemistry Department at the University of Delaware, which had close ties to DuPont; he also continued as a consultant.

Injected into the firm's departments at the rate of several dozen a year, some of its engineers worked on developing new production processes and new chemicals, while others were occupied with setting up, maintaining, and improving existing production processes. The work of the latter was rather similar to that of their predecessors, except that the adoption of unit operations now gave them a distinct professional identity, mirroring that of the mechanical engineers, who had adopted the Taylorite method somewhat earlier.

By the 1920s, the chemical engineers, who ten years earlier had been considered marginal at DuPont, began to arouse the sustained interest of the firm's leadership. There were two reasons for this change. One was the consolidation of a profession that had found in unit operations a concept that provided it with its own methodological identity and gave new chemical engineers the status of recognized specialists. The other was the diversification of DuPont, which, after timid beginnings in 1914, was implemented on a large scale after the war, when war profits made it possible and when the company realized that the key to its growth was a civilian market of mass-produced goods rather than its traditional and essentially limited specialty, the production of gunpowder and explosives. From this perspective, chemical engineers proved to be valuable assets when it came to diversifying the range of products and coordinating the synergies needed to deal with massive simultaneous investments in several fields. This was why DuPont was eager to establish ties to the universities, particularly MIT, and why its leaders felt that it was important to have proper coordination among teaching, research, and industrial work as such. This was not achieved without difficulties and without disagreements with MIT, especially with respect to MIT's general strategy in the 1920s. For this reason, MIT's change of scientific direction, negotiated in the early 1930s after Karl Compton was elected to its presidency, was bound to please DuPont's directors. But here too there was competition among departments when it came to hiring chemical engineers, who were coveted by many other companies as well. Recruitment procedures could, at times, be rather disorganized.

The company's manner of investing in its supply of young engineers was not without parallels to its efforts at vertical integration: it was necessary to hire suitable personnel, just as it was necessary to make sure that sources of regular supply of raw materials were available. But young engineers also found it advantageous to work for a large company, where the different stages of a successful career could unfold, and which, in their opinion, was also a source of considerable prestige. There was thus a triangular encounter between companies, universities, and young people who more or less by chance had specialized in chemical engineering, attracted to that field by what they read in brochures, by enterprising professors, and by a great deal of talk to the effect that times had changed, that chemical engineers were headed for brilliant careers, and that they would not be anonymous drudges of uncertain status. When the landlady of a chemical engineering student at MIT learned about his specialty, she exclaimed: "Oh, you are in chemical engineering! Do you realize that your future is assured?"[156]

In the early 1920s, a mechanism of mutual reinforcement between supply from the universities and demand from industry gradually came into play and channeled cohorts of students into chemical engineering, lured by bright prospects of rapid ascent into managerial positions. Indeed, their university training gradually came to include courses in management and marketing, which opened perspectives of access to a vast array of managerial positions. All of this further reinforced the distinction between chemical engineers and chemists and favored the professional advancement of the former.[157] The early inclusion of management courses in programs of chemical engineering was a sign that the universities had made the strategic decision to train future leaders of industry. The chemical engineers whom DuPont hired in the 1920s were the specifically American products of a carefully calibrated course of technical studies designed to respond to the new needs of industry. The era of the obscure, subaltern chemical engineer seemed to have ended.

CHAPTER TWO

From Ammonia to Nylon

Technologies and Careers

By the 1920s, chemical engineers may have been in a favorable position at Du-Pont, but this did not mean that they were sure to make a career there. A large company is a heterogeneous entity, made up of diverse and sometimes competing professional groups. In such a setting, professional authority based on the mastery of different kinds of technical and social skills does not last forever. The overall economic situation and the ups and downs of business also play a role in invalidating the idea of a permanent stabilization of professional groups. The chemical engineers had yet to prove their worth, as far as many of their colleagues were concerned.

What I intend to show here is how in directing the manufacture of ammonia, and then nylon, these engineers accumulated a capital of technical and symbolic authority that allowed them to consolidate their position in the company on the eve of World War II. In the case of nylon in particular, the engineers perfected a model of development that would, a few years later, transform what had started out as a marvel in the laboratory into a commercial product manufactured on a large scale.

There are many ways of writing the history of nylon. One can choose to analyze its technico-scientific origins or its marketing and commercial success. Here I have chosen to show how nylon completed DuPont's transformation from an explosives manufacturer that had branched out into production derived from explosives chemistry (such as paint and cellulose-based products) in the 1920s to a firm engaged in new fields where the know-how of chemical engineers proved relevant. The earlier and important stage that preceded nylon was the manufacture of synthetic ammonia in the 1920s and early 1930s, which brought DuPont into the technologically advanced domain of high-pressure chemistry.

Technical Cultures Within the Company

Decentralization of the Departments

In 1921, DuPont was the first major American company to adopt a multi-divisional structure. The industrial departments—there were ten of them in 1921—were now centered on their products rather than on their functions, whereas specific functions continued to define the service departments. Before the Great War, DuPont's activities had been essentially centered on a single line of products, gunpowder and explosives, not counting a few marginal activities. But then the implementation of the firm's new diversification strategy, pursued at breakneck speed once the war had begun, had created considerable difficulties.[1] Between 1915 and 1918, the number of employees rose from 5,300 to 85,000, and the capital from $83.5 to $309 million. More and more plants were added, and the line of products expanded excessively, ranging from artificial leather to paints, celluloid, and even an artificial fiber, rayon. Even though these products had certain features in common—above all the use of nitrocellulose as their basic ingredient—they were not all manufactured or managed in the same manner. Paint, for instance, was not sold like dynamite, through regional branch companies, but rather through a network of small independent dealers. DuPont's paint business was losing money, unlike some flourishing companies, such as Sherwin-Williams, the leader in this field, which sold its products to local wholesalers. Following lengthy debates among DuPont's principal managers, this was the reason for a change of direction in 1920, which centered the company's organization on products rather than functions. Alfred D. Chandler Jr. has shown that decentralization was intended to bring the firm's structure into line with its diversification, given the fact that its centralized structure and its functional departments no longer met the company's needs. The Paint and Varnish Department, for instance, was now given the freedom to choose the most appropriate commercial strategy for its purposes.[2]

Dupont was headed by its Executive Committee, which was responsible for its general strategy, and divided into financially autonomous product departments (see fig. 2.1). Each department was directed by a general manager charged with keeping the Executive Committee informed of his department's activities. The Executive Committee, for its part, did not have to concern itself with ordinary management tasks and could devote its efforts to the overall strategy of the firm. Even the central Department of Chemical Research was almost dismantled in December 1921, since each department was now responsible for its own research. In point of fact, various people within the firm had severely criticized this department for its inability to translate laboratory experiments to an industrial scale. Its budget fell from $3 million to $322,000 between 1920 and 1922, and its workforce was reduced from 300 to 21 persons.[3] With his usual frankness, Elmer Bolton, at the time director of chemical research in the Department of Dyes, commented that this was an opportunity for the firm to get rid of quite a few chemists who had been hired too hastily during the war and who had not proven their competence.[4]

This departmental decentralization brought positive commercial results, since most of the departments once again turned a profit, but it also had an unexpected consequence, namely, the preservation of cultures specific to each of the departments. The new structure tended to congeal their practices and their internal professional equilibrium. For the departments were now doing their own hiring, and their internal organization gave them ample room to maneuver. The product departments were not even obliged to work with the service departments if they could get a better deal elsewhere. To be sure, employees might change departments several times in the course of their careers, but interdepartmental transfers were fairly rare before World War II, except for upper-level managers. In a sense, the firm was divided into a number of independent companies.

The reform of 1921 must therefore be considered not only from the point of view of the executive committee—its internal conflicts and the changes it brought about[5]—but also from that of the departments, where it caused the hardening of certain tendencies and left other things unchanged. Here is a paradox that is not easy for a historian of organizations to deal with: how do

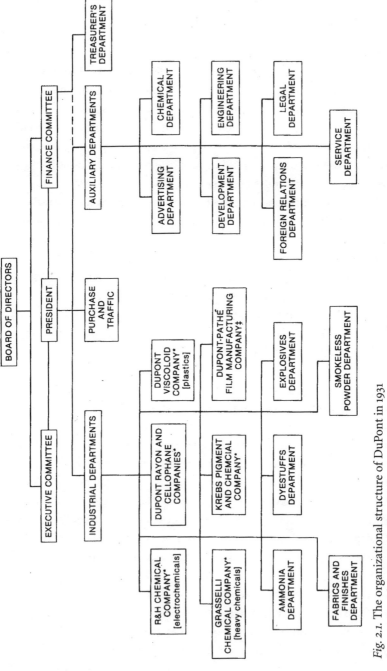

Fig. 2.1. The organizational structure of DuPont in 1931

Source: David A. Hounshell and John Kenly Smith Jr., *Science and Corporate Strategy: Du Pont R&D, 1902–1980* (New York: Cambridge University Press, 1988), fig. II.1 (p. 122), adapted from Du Pont Company Organization Charts, Imprints, Hagley Museum and Library, Wilmington, Delaware.

changes in overall structure limit change at the departmental level? The determining factor is the level at which the dynamic of change is analyzed.

At DuPont, three professional cultures coexisted until World War II (fig. 2.2). "Professional culture" is here taken to mean a cluster of knowledge and practices validated by the group and the organization to which it belongs, and which in turn creates a common and validating identity. During the interwar period, the older departments were still shaped by the culture of the shop floor (analyzed in Chapter 1), which was characterized by an inductive and semi-artisanal approach to the production processes. As we have learned, the gunpowder and explosives workshops were not places where chemical engineers found employment.

But what about the new departments issued from the firm's diversification? Two professional groups stood out: the chemists and the chemical engineers. By the 1930s, the development of nylon epitomized the success of the culture of chemistry and chemical engineering and marginalized the previously dominant workshop culture.

The Chemists' Departments

The first cluster of new departments was clearly different from Gunpowders and Explosives in the sense that they employed large numbers of university-trained chemists. Here, technical competence was based not so much on the tricks of the trade and long experience as on a thorough knowledge of chemistry and the mastery of complex processes in organic chemistry. In this category, one can include the Departments of Rayon, Paint and Lacquers, and Dyes, in other words, the departments that mass-produced consumer goods that sprang from organic cellulose chemistry, a specialty that made great strides in the 1920s. The job of developing organic chemistry products was usually assigned to chemists, with the understanding that they would improve products and processes purchased elsewhere. Here the chemical engineers had no more than a marginal and subordinate role that did not allow them to bring their own know-how to bear, particularly since the chemists were not inclined to give them carte blanche. Moreover, these departments of fine chemistry were partial to batch processing rather than to the continuous-flow processing championed by the engineers. Let us take a look at their best-selling products.

The Department of Dyes was essential to the company, for mastery of the chemistry of dyes carried over to the practice of organic chemistry as a whole (and indeed, it became the Department of Organic Chemistry in 1931). Until

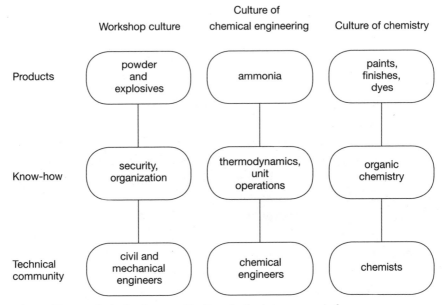

Fig. 2.2. Three technical cultures at Du Pont in the interwar period

the early 1930s, this department was in a rather difficult situation, for DuPont had not entirely mastered certain aspects of organic chemistry. Bolton, at the time director of research for this department, started out with the modest objective of reproducing the German dyes, an objective that was favored by very high customs barriers against the importation of dyes under the Fordney-McCumber Tariff Act of 1922.[6] But the task was difficult, for, as Bolton pointed out, the German patents for dyes were usually written in an extremely vague, even fallacious manner, which meant a great deal of extra work for his chemists. But he later conceded that in addition, these men were too young and inexperienced and lacked the requisite know-how.[7] The situation eventually improved, the chemists mastered a certain number of processes, and new dyestuffs were produced. Like other fine chemicals, dyes were produced in batches under the direction of chemists.

Rayon, a new product that played an important role in DuPont's diversification in the civilian market, is another example. At the time, it was widely used by fashion designers, and its reasonable price—one-third that of silk—made it accessible to the middle classes. The Rayon Department was heir to the very first efforts at diversification DuPont had undertaken early in the century, when it had set out to find new uses for cellulose. As early as 1909, the chemist Fin Sparre had been sent to Europe to study the manufacture of cellulose products. At that time, he had become particularly interested in "artificial silk." DuPont eventually acquired the manufacturing process and the product from a French firm, the Comptoir des textiles artificiels, and the first manufacturing plant it built at Buffalo was an exact replica of the French plants.[8]

The Rayon Department included several doctors of chemistry, who were charged with developing a proper viscose solution and processes for manufacturing thread. The difficulty here was to obtain a continuous thread of uniform quality from the viscose solution.[9] The same problem was to arise a few years later for nylon. One chemist had the idea of passing the solution through a metal shield perforated by tiny holes, the so-called spinnerette. The filaments were then immersed in a bath of sulfuric acid in order to solidify them. For this purpose, the chemists constructed a series of machines that made it possible to obtain uniform quality. In his study of the DuPont rayon factories, the economist Samuel Hollander has demonstrated that these gradual improvements actually produced the greatest gains in productivity.[10] If the manufacture of rayon brought a special know-how to the field of fibers that was to prove invaluable when it came to developing nylon, it was primarily derived from the work of chemists.

The other leading product of the Rayon Department was cellophane. Its patent and its manufacturing process had been acquired from the Comptoir des textiles artificiels, from which rayon had also been bought, and French technicians came to install the production unit in 1924.[11] Since cellophane had the same chemical composition as rayon, the difficulty here was to transform the viscose material into a transparent film rather than into a thread. A young chemist in the department, Bill Charch, improved the cellophane by making it resistant to humidity, so that it could be used for wrapping food products. The department's objective was to improve the production processes by combining chemical experimentation with mechanical tinkering.

All in all, assessing DuPont's chemical capabilities until the end of the 1920s calls for nuanced judgment. In most cases, one sees more or less clear-cut improvements of manufacturing processes and of products purchased elsewhere, and above all good management and good commercial strategies. DuPont was the most high-powered chemical firm in the United States primarily because of its organizational capabilities and its ability to take on very large-scale projects. Moreover, it was highly skilled in protecting its products and its processes by means of a wall of patents drawn up by a team of lawyers (who worked together in an autonomous service department, a novelty for that era). And it knew how to market its products efficiently, once it had analyzed the problems it had faced in the early 1920s. Nonetheless, DuPont in that era is "a notable example of a major laboratory's inability to invent and develop processes and products in which the company and the laboratory had no substantial experience," Thomas Hughes observes (contrary to David Hounshell, who in my opinion overestimates DuPont's scientific capabilities at the time).[12]

However, in some sense, this controversy is beside the point: what was really at stake in these departments was finding the hinge between the purely chemical work and its translation to an industrial scale. Partly a matter of mechanics and partly one of chemistry, this raised the question of how a chemist like Bolton was to work with his engineering colleagues. By and large, DuPont's departments maintained the chemical industry's traditional separation between the chemists—who predominated—and the engineers. If they did employ engineers, these were for the most part production engineers working under the direction of chemists, as in the case of Robert MacMullin. Bolton insisted that the chemist who had developed a manufacturing process must monitor its industrial stages and be placed in charge of this production.[13] The only case of a chemical engineer developing a production process was that of tetraethyl lead [used as an anti-knock agent in gasoline until it was phased out in the 1970s as a deadly poison], and so Bolton entrusted this production to William Calcott, a young engineer and graduate of the University of Virginia. Significantly, Bolton made the point that Calcott was "a chemical engineer, but also a very good chemist," which was his way of signaling that Calcott was indeed an exception, and that the chemists predominated in his department.[14] It should be added that the chemical engineers at DuPont always resented Bolton, an attitude that clearly shows in their interviews.

By contrast, two other departments, Ammonia and Chemical Research, constructed a different culture, which I call the "culture of chemical engineering." These departments had no connection with either the culture of organic chemistry or the Gunpowder and Explosives Departments's shop-floor culture. They had their own engineering culture, a twentieth-century phenomenon represented by the new generation of chemical engineers who had been trained in the 1920s at MIT and a few other universities.

Synthetic Ammonia and High-Pressure Chemistry

The Haber Process

The creation of synthetic ammonia was certainly one of the important moments in the history of modern industry. But it is a moment that has not commanded a great deal of attention on the part of historians, no doubt because the history of industrial chemistry remains a poor relation of the history of industry, and also because high-pressure chemistry, of which ammonia production is a part, has overshadowed it.

Ammonia (NH_3), which is made from nitrogen (N) and hydrogen (H), is a basic product needed for the manufacture of nitrogen-based fertilizer—as the German chemist Justus von Liebig had already demonstrated by the middle of the nineteenth century—and of explosives.[15] By the late nineteenth century, the search for a convenient, economical process for manufacturing ammonia occupied the best minds in European chemistry. In light of the heightened rivalries among the major European powers and their demographic growth, ammonia was considered a crucial asset in meeting the needs of armies and populations. In a sense, synthetic ammonia was to invalidate Malthus's somber predictions, while at the same time opening new military perspectives.

The most frequently used process for manufacturing ammonia consisted of treating natural nitrates from Chile. But these reserves, while abundant, were not inexhaustible, as experts were beginning to realize at the time.[16] The remaining needs for nitrate were covered by ammonia sulfate obtained through the distillation of coal in coking and gas plants. To be sure, the methods using cyanamide or the arc that had been developed in Germany and in Norway made it possible to obtain synthetic ammonia, but their yield was low and their electricity consumption very high.

The chemists actually knew that the air contains a high proportion of nitrogen (80%), but until the late nineteenth century, no one knew how to fix the nitrogen for the purpose of converting it into ammonia or nitric acid. An important first stage was therefore the liquefaction of air, achieved by the German chemist Carl von Linde in 1898, which made it possible to obtain nitrate by distillation. His method was used by the French chemist Georges Claude, who improved the process and in 1902 founded the Société d'Air liquide to exploit it (1902).

Better yet, in 1908 a chemist and a metallurgical engineer of BASF, Fritz

Haber and Carl Bosch, succeeded after several years of research in making synthetic ammonia, using the following formula:

$$N_2 + 3H_2 \leftrightarrow 2NH_3 \text{ (+ heat)}.$$

By combining nitrate with hydrogen in a pressure vessel with a catalyst, they obtained ammonia that was easy to liquefy and could be used to produce nitrogen-based fertilizers or nitric acid. It had now become possible, as a famous expression had it, to "make bread out of air."

At the time when Haber and Bosch developed their process, the chemical industry already had a certain amount of know-how in high-pressure chemistry. One of the first pieces of high-pressure equipment had been invented by Denis Papin in 1681 for the purpose of extracting marrow from bones. Having noticed that this was easier under pressure, he built a bronze autoclave equipped with a safety valve. But it was really only in the middle of the nineteenth century and in connection with the development of artificial dyes that autoclaves, usually made of cast iron, came into widespread use in chemical factories.[17] Pressures of from 30 to 100 atmospheres made it possible to produce the amines and other aromatic bases for the dyestuffs that enriched German chemical firms.

However, the production of synthetic ammonia required considerably higher pressures (200 to 250 atmospheres).[18] Generally speaking, high pressure accelerates chemical reactions by rapidly breaking the connections between atoms. Haber's setup combined the use of a catalyst with high pressure in a process that, once it was adapted to a large scale, made the fortune of BASF. It also made the fortune of Haber, who was appointed head of the prestigious Kaiser Wilhelm Institute of Physical Chemistry in 1912 and in 1919 received the Nobel Prize in recognition of his work with ammonia. This prize was controversial, however, for the Allies objected on the grounds that the German scientist had developed poison gases during the war. Bosch also obtained a Nobel Prize a few years later, in 1931, when he was chairman of IG Farben.[19]

The synthesizing of ammonia was unquestionably a decisive moment in the history of industrial chemistry. In terms of chemistry, it was much simpler than certain processes in organic chemistry, but this was not the case when it came to the constraints of the metallurgy. It took several years of testing to develop sufficiently solid, airtight compressors. Manufacturing ammonia required the use of compressors and conduits capable of withstanding high temperatures and very strong pressures, so that it became necessary to develop special kinds of steel. Hydrogen under high pressure is very corrosive of steel, and metals are weakened by high temperatures. Hence the need for new valves, new joints, and more precise and solid measuring devices than in the past, for they had to be capable of measuring the temperature, the composition, and the density of the gases and their degree of oxygen and hydrogen concentration. The compressors (where the hydrogen and the nitrogen were mixed) for BASF's first ammonia plant at Oppau weighed 65 tons each, and their walls of high-grade Krupp steel were one meter thick. The plant included, upstream from the synthesis, facilities for producing hydrogen from water gas and liquefying the nitrogen, while downstream the ammonia was transformed into nitric acid.

The cost of operating an ammonia plant is fairly low, since it essentially uses coal or petroleum as its raw material, but on the other hand, considerable capital must be tied up, given the construction costs: economies of scale are thus a major factor, somewhat as they are in the nuclear industry.

During the 1914–18 war, BASF increased its production capacity for ammonia to 160,000 tons per year, thanks to the construction of the Leuna plants. This allowed Germany to produce the smokeless gunpowder and explosives that its armies needed. Victor Cambon, who visited Leuna in the early 1920s evoked "thirteen gigantic chimneys, spaced about 200 meters from each other, that continually belched white smoke."[20] In his third book about Germany, Cambon actually devoted a whole chapter to Germany's "nitrogen policy" and several pages to the Haber process as described above at the Oppau site, "which our air force, out of inexplicable negligence, failed to destroy":

Nothing in any other industry is as impressive as these cyclopean constructions, these huge, strangely shaped apparatuses, these forests of tubes of all calibers that rise up, descend, and crisscross above our heads, these bundles of electric cables, these networks of aerial conveyors that carry wagonloads of coke in every direction 25 meters above the ground, these cranes, these elevators, these puffing compressors, these silos, this battery of 18 cylindrical towers, 30 meters high and 8 meters in diameter, lined with materials impervious to acids and enclosed from top to bottom in a covering of sheet metal.[21]

Without the Haber process, the armies of the Reich might not have been able to fight until 1918. If one adds to this Haber's role in chemical warfare, one understands the importance of this scientist, who was engaged in both theoretical and experimental work in close collaboration with a private firm

(BASF) and the German state. A German, a Jew, a nationalist showered with honors, and a wealthy man (he received a pfennig in royalties for each kilo of ammonia produced), Haber proposed scientific-technical solutions—to be implemented under his direction—to industrialists and to the military. He had known how to make himself indispensable to everyone. But for all that, the celebrated chemist had to take hasty leave of the chairmanship of the Kaiser Wilhelm Institute for Physical Chemistry when the Nazis came to power in 1933 (he died in Switzerland the following year). Today one can read on a stela erected to his memory in Berlin in 1952 that "Themistocles is known in history not as the exile at the court of the king of Persia but as the victor of Salamis," which is a way to suggest that Haber should be remembered as the inventor of synthetic ammonia rather than of mustard gas. But the two products are connected, for both were used for military and commercial objectives that Haber himself presented as one and the same in the name of making Germany great.

DuPont's Early Ammonia Production

Under the Treaty of Versailles the French, the British, and the Americans were allowed to inspect the German ammonia plants, to copy their designs, and to interrogate the German technicians.[22] The representatives of the victorious countries took full advantage of this opportunity, and ammonia plants were soon under construction in such places as Billingham, England, and Toulouse, France.

For the Americans, the need for synthetic ammonia was less critical, for they could call on natural reserves of nitrate. DuPont in particular had taken the precaution of buying enormous quantities of Chilean nitrates, most of which were stocked and managed on the premises by a small group of DuPont employees, among them the young Walter Carpenter, who was to rise through the firm's ranks all the way to the chairmanship.[23] The U.S. government belatedly commissioned facilities at Sheffield and Muscle Shoals in Alabama using the cyanamid process, which turned out to be rather inefficient and became controversial because of their cost.[24]

After the war, however, DuPont did become interested in synthetic ammonia. Its coffers were full, so it could afford large-scale investments. In particular, the company intended to pursue its diversification into the fields of paints, dyes, and other chemical products that used nitric acid or ammonia, often in combination with cellulose. Its leaders were well aware that the Chilean reserves were not inexhaustible, but they were also intent on acquiring an expertise in high-pressure chemistry, whose importance had been demonstrated by BASF. At this point, all of the world's major chemical companies were keenly interested in this new technique, and DuPont did not intend to be left behind. Moreover, certain members of Congress favored the development of synthetic ammonia for reasons of independence and national security.

As soon as the Armistice was signed, DuPont assembled a rather disparate set of young engineers to start in on high-pressure chemistry. Among them were four former employees of the Fixed Nitrogen Laboratory of the U.S. Department of Agriculture, some recent engineering graduates, others who had had some earlier experience with synthetic ammonia in government-run experimental plants, and also some French technicians. Heterogeneity was to be the strength of this group of young people, who perfectly exemplified the intense exchanges that established ties between public and private institutions of management, knowledge, and business. We shall return to these later.

DuPont's road to producing synthetic ammonia was rather tortuous and fraught with pitfalls. Instead of choosing the Haber process, relatively well-tested though it was, DuPont opted in July 1924 for a new and theoretically promising process developed by Georges Claude of Air liquide. The Claude process consisted of using extremely high pressures, close to 1,000 atmospheres—or five times higher than those of the Haber process—which in principle made it possible to obtain five times as much ammonia from the same quantity of nitrate and hydrogen. At the same time that DuPont signed an agreement with Air liquide, it also acquired the only American company that experimentally produced ammonia (using the Haber process), the National Ammonia Company.[25] In August 1924, the three partners were consolidated in a holding company, Lazote Incorporated, which was legally independent of DuPont, being controlled by three different partners, As it had done for cellulose, DuPont entered the field of ammonia by calling on human and technical resources outside the firm. This time, however, an affiliated yet distinct organization was created, no doubt because in the case of ammonia, the technological threshold was much higher than for cellulose, and because DuPont itself did not have the necessary know-how.

Construction of the plant began in May 1925 at Belle, West Virginia, close to the Appalachian coal fields. It included facilities where in a first stage, pure hydrogen could be produced. Hydrogen was obtained by burning coal and recovering a gas (carbon monoxide and hydrogen) from which the carbon monoxide was subsequently removed. After the hydrogen had been mixed with nitrate—at the rate of three parts hydrogen to one part nitrate—it was placed under pressure as a way to obtain ammonia. In order to compress the hydrogen-nitrate mixture, very large compressors, the so-called hypers, were used, but they often broke down. The pressures of the Claude process were very high indeed. The plant was constantly paralyzed by technical problems related to pressures and temperatures that overtaxed the machinery and caused sometimes fatal explosions.[26] Accidents were not rare in the ammonia industry as a whole: in Germany, the Oppau plant blew up in September 1921, killing 600 workers. At the end of the 1920s, Belle was still an experimental establishment, whose processes had not yet been worked out. Unfortunately, documents about this plant are lacking, and it is difficult to further explore the conditions under which experimentation was carried out there and to describe the work done at the Belle facilities.

What we do know is that by 1928, the increasingly ferocious competition among chemical firms in the ammonia market resulted in lower prices. The great industrial powers now produced ammonia in quantity, and their production capacities were twice as high as worldwide consumption (3 million tons produced against 1.5 million consumed).[27] In the United States, DuPont was no longer alone in the ammonia market: Mathieson Alkali Works, Roessler and Hasslacher, which operated near the Niagara Falls, Midland Ammonia Company in Michigan, and the Great Western Electro-Chemical Company in California were serious competitors who had mastered the techniques of high-pressure chemistry.[28]

Then, in 1929, DuPont bought up its associates' share in Lazote Inc. and proceeded to set up the DuPont Ammonia Corporation, which in September 1931 became the Ammonia Department, a department entirely devoted to industrial production.[29] But the Great Depression only aggravated this sector's stagnation. At the height of the crisis, in 1931, the situation of the Ammonia Department was precarious indeed. DuPont had invested considerable sums, first in the Claude process, then in another, related process, the Casale process, and finally in the Haber process; it had also built large production units that had never operated at full capacity.[30] In the space of some ten years, DuPont had spent about $30 million—half of it on Belle—which was almost half of its net earnings from World War I, amounting to $68 million.[31] It should be noted in passing that IG Farben was in a similar situation, and that it was the Nazi seizure of power that saved the BASF division from bankruptcy, for thereafter the German government subsidized high-pressure facilities for producing synthetic gasoline in preparation for war. Moreover, as the historian Ludwig Haber has pointed out, the demand for ammonia—mainly a function of the needs of farmers—is highly seasonal, so that the ammonia produced in a steady stream throughout the year must be stored in huge storage facilities.[32]

Why did DuPont make these heavy and initially rather unprofitable investments? While we do not have the minutes of the Board of Directors' discussions about ammonia, a certain number of documents from Lazote Inc. and the Department of Development have been preserved.[33] At the time of the original investment in 1924, Fin Sparre argued that it was necessary both in order to obtain sufficient ammonia without depending on imports from Chile and to acquire know-how in high-pressure chemistry. Later, when the price of ammonia collapsed, the accumulation of know-how became paramount. Throughout the 1930s, the DuPont directors devoted many reports and discussions to the subject of ammonia, the investments earmarked for its production, and future plans for development. The investments made over the years committed the firm to a track from which it became increasingly difficult to exit. These weighty scientific and technological choices preoccupied and concerned the organization.

The consensus that gradually took shape was summarized by Walter Carpenter in a note of October 1939, in which he pointed out that the future of the Ammonia Department had not been compromised, despite the dismal state of the ammonia market. Indeed, Carpenter asserted that "at the moment . . . we should direct our efforts and our investment in the future toward the utilization of our technique, our man-power, our money and our basic products from the ammonia industry in the development of new industries and new products . . . rather than using those efforts in the further enlargement of capacities in the highly competitive fields."[34]

Was Carpenter trying to rationalize the investment in ammonia retroactively, at a time when the department had weathered the crisis? However that may be, in its ammonia business DuPont softened its usual criteria of financial profitability in favor of other considerations. In this case, its directors approved investments intended to accumulate technical and human capital and did not adhere to strictly financial criteria only. The case of ammonia clearly shows that DuPont's management model was not as rigidly focused on financial profitability as studies of its management have maintained.

Engineers Under High Pressure

These considerable and, at least in the short run, rather unprofitable investments were to prove decisive in the renewal of the company's human and technical capital. That capital crystallized around two of DuPont's departments,
the Ammonia Department, the offshoot of L'Azote, Inc. (the joint venture with
Air liquide), directed by Roger Williams; and the Department of Chemical
Research, or more precisely the chemical engineers of this department, who
in 1929 were given a formal organization and placed under the direction of
Thomas Chilton. To this category one must add a satellite group of engineers
of the Engineering Department. An examination of these two groups and their
activities provides a good illustration of the characteristic features of an American high-tech company in the interwar period.

Roger Williams was born in 1890 in Nebraska into a rather modest family
recently arrived from Pennsylvania.[35] A good student with a talent for science,
he studied chemistry at the University of Nebraska (1909–11) before leaving
for MIT, where he obtained his B.S. degree in physical chemistry in 1914. He
then embarked on a doctoral program as a student of Arthur Noyes, but left
MIT prematurely in 1916, apparently in consequence of his marriage, and went
to work for the Nitrogen Products Company of Providence, Rhode Island, a
small company that was developing its own process for making nitrogen. One
wonders whether Williams might perhaps have become an academic if he had
obtained his doctorate. In any case, it is clear that a young chemist without
a doctor's degree could not aspire to a university career, and that businesses
outdid the universities in the pursuit of the best scientists whenever the latter
did not have the most prestigious degrees. Two years later, Williams came to
DuPont as research supervisor in the Department of Chemical Research. At
that time, and despite his youth, his qualities as an organizer and leader of men
already seem to have been evident. He was therefore promoted, as were many
other of the firm's young engineers and managers.

Williams was convinced that the best process for producing ammonia was
Haber's, and it was thus against his advice that DuPont chose the Claude process.[36] In 1924, he was appointed technical director of L'Azote, Inc., and later of
its successor, the Ammonia Department. This was a strategic function, given
the uncertainties about which process to use and the great complexity of high-
pressure chemistry. Williams was to occupy this position until 1942, the year

when he was called to a different function, connected with the Manhattan Project. The Ammonia Department's research budget amounted to 7 percent of the department's sales revenue in 1929, and rose to 10 percent in 1930, which was three times more than the usual and average percentage for the company as a whole.[37]

The tightly knit team of chemists and chemical engineers that Williams assembled was capable of building and managing some of the most advanced industrial equipment of that time. After all, they faced not only complex theoretical problems of physics, but also a variety of practical problems, particularly those concerning the strength of materials.

Williams headed a team of which he had personally chosen each member. R. M. Evans was the production manager; John Woodhouse, responsible for the department's sales, was a chemist who had obtained a Ph.D. at Harvard in 1927 before joining DuPont the following year;[38] and Charles Cooper had been director of one of the workshops at MIT's School of Chemical Engineering Practice from 1928 to 1936. Between 1936 and 1940, Cooper, a pure product of MIT and an experienced practical engineer, was employed at the Belle plant, working on improvements for its production processes and on the manufacture of intermediary chemicals needed for nylon.[39] Evans, Woodhouse, and Cooper were to join Williams during World War II. These young engineers had little in common with the old hands of gunpowder-making. As Chaplin Tyler puts it: "Du Pont was still manned mostly by old-time powder people, and this field of high-pressure catalytic gas technology was completely foreign to these old-timers. I don't think they had any interest in it, much less appreciation of what it was about."[40]

Woodhouse remembered the disdain for engineers shown by Clark Davis, a manager in the Graselli Chemicals Department and former employee in the Explosives Department.[41] By contrast, everyone had the greatest admiration for Williams. Woodhouse asserts that he went to work for DuPont because he had met Williams, and so did most of his colleagues.[42]

Yet at that point—during the worst years of the Great Depression—the problem for Williams and his team was to apply their know-how in high-pressure chemistry to other areas than the manufacture of ammonia and nitric acid, whose prices had collapsed. The price for a pound of ammonia had fallen from 32 cents in 1928 to 16 cents in 1932. Hence the search for diversified production at the Belle plant. The head of the Ammonia Department tried—unsuccessfully—to sell ammonia as a coolant for refrigerators.[43] As the departmental managers were frantically looking for customers within the firm, they had to convince the executive committee that such action was necessary. Ammonia was more than just another product, it was also a flood of talk designed to convince and justify; it was words and diagrams produced in interminable meetings, where Williams learned the art of persuasion. All of this was supplemented by invitations to visit the Belle plant, whose facilities impressed the visitors, among them politicians such as Nebraska Senator George Norris, the future father of the Tennessee Valley Authority, who came away in amazement. Innovation is a matter not only of products, but of practices as well, including the skillful handling of language. Objects are coated with discourse, with permanent negotiations that will bring them to safe harbors.

A highly profitable product was methanol (obtained by combining carbon monoxide and hydrogen under pressure and then changing the catalyst), which was notably used as an antifreeze called "Zerone."[44] The Ammonia Department also manufactured other alcohol derivatives, such as paint thinners and brake fluid for Lockheed, synthetic urea for fertilizers (obtained by mixing ammonia with carbon dioxide), and even a commercially successful plastic resin, lucite (produced from ammonia, methanol, and carbon monoxide).

But this was not sufficient to use the expensive industrial equipment of the Belle plant to full capacity. Even after $24 million had been invested between 1924 and 1934, the Ammonia Department was still in the red. It was sixth among DuPont's eleven departments in terms of investments but next to last in profitability.[45] In 1934, the Belle plant represented a capital investment of $50 million, a record at DuPont. Something else had to be found. Williams's offices happened to be located close to those of the chemists in the Department of Chemical Research, and Williams carefully watched the research conducted there, for the future of his department depended on it. Did he receive explicit warnings from the Executive Committee? Woodhouse believes he did, for he remembers that in 1933 Williams spoke to him about closing up shop before too long if the situation did not improve.[46] Woodhouse remarked that "we had no standing, no expertness in the eyes of other departments. Central Research and Orchem had the standing . . . so it was [them] against this upstart department which had lost money for the company from the start. . . . It was touch and go."[47] Tyler adds that DuPont was not in the habit of losing that much money, but that he said, "we are all in the ship together, as I see it. Are you going to sink with me or are we going to try a little swimming?"[48] These engineers may have been dramatizing their situation a posteriori, the better to highlight their future accomplishments. The Executive Committee did not consider shutting down the Ammonia Department, even if it was clearly in difficulty.

The ammonia engineers had neighbors working in the Department of Chemical Research under the direction of Thomas Chilton. This group of some ten young engineers had been formally assembled on 1 July 1929. At the time, it was a subdivision of the Department of Chemical Research charged with doing research in chemical engineering with a view to gaining a better understanding of chemical reactions and adapting the equipment accordingly. On 13 May 1935, this group of chemical engineers was transferred to the Department of Engineering, where it became its technical division in charge of designing and building all plants. At the time, the director of this division was Henry B. du Pont, with Thomas Chilton as his associate (he was to succeed du Pont in 1938). Some of these engineers formally remained members of the Department of Chemical Research.[49]

The offices of these engineers were located in the same building at the Experimental Station. Behind the building were a few acres of lawn, where the employees could relax or have lunch at noon. The engineer Crawford Greenewalt, who later became DuPont's president, would sometimes sit there to practice his clarinet, much to his colleagues' distress.[50] Nearby was Purity Hall, the facility especially built for the chemists in "pure" research who had been hired since 1927. The geographic and intellectual proximity of the engineers had not come about by chance. The engineers in chemical research could point out to their superiors that their activities were useful to the ammonia and methanol specialists. The latter, for their part, benefited from the experiments with high pressure carried out by the engineers of the neighboring department. And if they should have to work with recently hired chemists, being close to the laboratories would also be an advantage.

The reorganization of 1935 brought little change to the functioning of this small group of engineers, which did not even move from the Experimental Station. Aside from Chilton, it included Crawford Greenewalt, Art Larchar, Hood Worthington, Howard Young, Raymond Generaux, Thomas Drew, and Allan Colburn, who were joined by a few other engineers in the 1930s. These men were of the same age: Chilton and Worthington were born in 1899, Generaux, Greenewalt, and Drew in 1902, and Colburn in 1904. All of them were to stay at DuPont and work for the Manhattan Project during the war, except for Colburn, who left to become head of the Department of Chemical Engineering at the University of Delaware, and Tom Drew, who went to teach at Columbia

University in 1940. Between 1943 and 1945, however, Columbia lent Drew to DuPont so that he could work on the Manhattan Project before returning to Columbia as head of the Chemical Engineering Department, consultant for nuclear questions, and chairman of the visiting committee for the Brookhaven National Laboratory. Many of these engineers had known one another since their student days, for Chilton and Generaux had been students at Columbia, while Greenewalt and Drew were classmates at MIT.[51]

Among these young engineers, Greenewalt was a characteristic but also unique figure. He was born in 1902 in Philadelphia into a well-to-do family related by marriage to the du Ponts.[52] His father was a physician; his mother, a concert pianist. His maternal aunt Ethel Hallock had married Pierre, Lammot, and Irénée du Pont's brother William K. du Pont, and Crawford Greenewalt himself married his childhood friend Margaretta du Pont, a daughter of Irénée's, in 1926. Although this family connection did not slow down his career at DuPont—quite the contrary—it was not a decisive factor in his promotion, but two of the elements of Greenewalt's professional success were nonetheless his relationship to the du Ponts and his patrician background, with its characteristic perks, such as the trip to Europe the young man took with his mother, "a classic thing to do [for members of the Eastern elite]" as Greenewalt put it.[53] Later, his colleagues at DuPont noticed immediately that he had "connections," as Woodhouse put it. This privileged background contrasts with the more modest social origins of most of the other chemical engineers of the group, who generally came from the lower middle classes and improved their social standing by becoming engineers. Roger Williams, for example, who might also have aspired to the chairmanship of DuPont by reason of his great prestige and the appreciation of his colleagues, was a discreet man who had no use for a fancy social life: self-effacing to the point of appearing dull, he was never able to climb the last rung of the management ladder.

Greenewalt entered MIT at the age of sixteen, a month after the Armistice. As he remembered it, his decision to specialize in chemical engineering was probably influenced by the fact that he might some day be able to work for DuPont and by the personalities of the professors of chemical engineering at MIT, in particular, Warren Lewis, one of the co-authors of *Principles of Chemical Engineering,* a charismatic teacher who attracted many students to the field. Greenewalt obtained his bachelors' degree in 1922, at a time when very few students pursued graduate studies in chemical engineering; MIT awarded the first doctorate in this field in 1923. Greenewalt frankly admitted that "the

Du Pont Company would never have employed me at the Experimental Station if I'd come there ten years earlier, because they would have insisted on a Ph.D. But the whole atmosphere of a scientific education was a very different thing then."[54] In his account, Greenewalt seeks to make it quite clear that the modesty of his degree was only apparent, and he also stresses his youth (he was two years younger than his classmates). This was his way of showing that there was a firm and coherent relation between his educational attainments, his professional position, and his career, unlike in the case of another son of the Philadelphia Quaker elite, Frederick Taylor, the self-taught engineer par excellence.

Greenewalt—only twenty years old—went to work for DuPont in the fall of 1922 as a newly minted MIT graduate. Assigned to Philadelphia's Harrison Brothers plant, where paints and basic chemicals were produced, he was put in charge of product inspection. After he had worked for a few months in the paint and solvents factory, he quickly made a name for himself by improving a production process for sulfuric acid, for which he received an "A-bonus," a reward bestowed on deserving employees. Greenewalt noted that at the time he, like all other employees, had only "two weeks of paid vacation," and that the work was shared by two teams, the day shift (working eleven hours) and the night shift (working thirteen hours). By this he sought to convey that he was in the same boat as his colleagues, including the workers, and that his background and his degree did not in any way exempt him from the general constraints of the factory. A 1951 *Time* magazine article devoted primarily to him indicates that at that point his salary was only $120 a month, the equivalent of a skilled worker's wages. "Greenewalt was a wage earner like everyone else" working on the manufacturing processes for sulfuric acid.[55]

But before very long the mechanisms of socioprofessional selection came into play in the young man's favor. A first move brought him from his workplace in the factory to the research laboratory, located near DuPont's headquarters, and this geographic move closer to the organization's center of decision-making was another favorable sign. The success of Greenewalt's work had attracted the attention of Charles Reese, the director of chemical research, who interviewed him. "I suppose he thought I would be alright since he said, 'Come down here [to Wilmington],'" Greenewalt recalled.[56]

And so Greenewalt moved to Wilmington in October 1924, just when Reese handed over the department to Charles Stine. At that point, the Department of Chemical Research was barely managing to keep going, for the reorganization of 1921 had reduced its activities by distributing research and development activities among all of the company's industrial departments. After a few months during which "nobody quite knew what to do with [him]," Greenewalt's first real task was to design and build a high-pressure laboratory. This is quite understandable if one remembers that just then DuPont was experiencing difficulties with the Claude process for ammonia production. So Greenewalt tinkered to come up with a hydraulic pump with which to inject chemicals under high pressure into heavy steel cylinders. The objective was to devise reactions that would yield other products than only ammonia and methanol, as a way to amortize the high-pressure machinery. "The high pressure technique that we had was extremely rudimentary," Greenewalt remembered. The engineers of the Experimental Station were few in number, and sometimes they felt isolated in a laboratory totally focused on organic chemistry, to the point where Greenewalt seems to have worried about the future of his career:

> [I]t worried me a great deal from the point of view of whether I could ever advance in the organization, as a chemical engineer. I thought quite often maybe I'd just have to get out and get in some other area, which I didn't want to do because I love research. And this was particularly true of Bolton's time [when he was] heavily oriented toward organic work. When the central Research Department got up to a hundred technically trained people, I don't suppose there were more than five chemical engineers in the bunch. The thing was so heavily oriented toward organic chemistry that I wondered whether I could ever get along.

Greenewalt insists on his temporary troubles as a way to emphasize that he had to convince the chemists that he was doing good work. He evokes one of his efforts to improve the manufacturing process for lead tetraethyl. "Doc Chambers (a chemist) told me: 'No, it can't be done, it's impossible.' I said: 'I've done it.'" In the 1920s, the chemical engineers were still looked upon with some condescension by the chemists and had to prove their competence almost on a daily basis. Bolton in particular seems to have had little use for them at that time. "Bolton was an organic chemist, and I won't say that if it wasn't organic chemistry it smelled, in his point of view, but it wasn't very far from that."[57] The company's directors had made chemical engineering an integral part of their strategy for expansion and diversification, but the role and the status the chemical engineers would eventually assume within the company was by no means clear yet, particularly in the Department of Chemical Research. Hence the initial difficulties encountered by Greenewalt and the other young engineers in this department, and their initial feeling that "nobody quite knew what to do with them." The situation was rather fluid, and a chemist like Bolton had not yet understood the potential value of the newcomers. As late as 1942, Greenewalt, by that time head of research in the Grasselli Department, seems to have been disillusioned about his career prospects at DuPont.[58]

The chemical engineers intended to reinforce the technological identity of chemical engineering by providing the unit operations with the most theoretical and mathematical foundations possible in order to anchor their position within the company and expand the range of chemicals produced under high pressure. The problems associated with high-pressure chemistry required that a single theoretical approach be shared by the Departments of Ammonia and of Chemical Research, as well as by the design engineers in the Engineering Department. Consequently, the chemical engineers of the Department of Chemical Research actually spent most of their time on questions concerning the Ammonia Department. They appear to have had a certain margin of autonomy: "In the beginning the chemists had little interest in what we were doing," Greenewalt confided.[59] If he and his colleagues were mainly working on high pressure, it was no doubt because the most promising possibilities for chemical engineering lay in this area.

The question of horizontal coalitions has been analyzed by the management scholar Henry Mintzberg, who speaks of a "political system" within the organization in this connection.[60] In the present case, it took the form of alliances that, though not formalized in any organization chart, were concluded between groups or individuals with a view to increasing their power.[61] When one reads their studies and observes their way of organizing, it becomes quite clear that these young engineers were working in a concerted manner to promote their methods of work and to strengthen their positions within the firm. Here, taking a cue from Michel Winok's concept of "communities of ideological systems," one might perhaps speak of a "technical system community" at DuPont (fig. 2.3), meaning a generational effect reinforced by a shared approach to technical processes that stoked the fires of high-pressure chemistry.[62]

The objective of the chemical engineers in the Department of Chemical Research was to strengthen the scientific foundations of chemical engineering. This was a matter of creating mathematical models for certain unit operations in order to adapt them to the requirements of high-pressure chemistry. More was involved here than only "technique," for in reinforcing the theoreti-

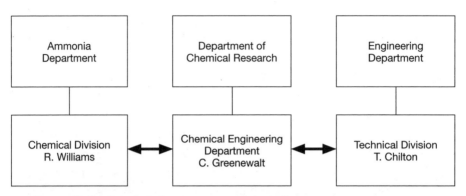

Fig. 2.3. DuPont's technical systems community in the 1930s

cal modelization of the discipline, the engineers also reinforced their own differential position in relation to the engineers of the older generation; above all, they created a production tool that combined techniques of mass production (as in the case of traditional heavy chemicals) with those for high value-added products. What was new about this approach? Here we must open the dossier of the technological evolution of chemical engineering.

The two principal questions to occupy the engineers of these two departments were, on the one hand, high-pressure chemistry—both for lowering the production cost for nitric acid and for manufacturing ammonia and methanol—and, on the other, the more fundamental matter of the behavior of liquids and gases in the conduits and reactors of the chemical plant.

The engineer can examine these phenomena of the behavior of liquids and gases in several different ways. The concept of the unit operations, as formalized at MIT by Walker, Lewis, and MacAdams, provided an appropriate starting point, particularly the notions of fluid, heat transfer, and mass transfer. These notions were more fruitful than those of "stirring," "grinding," or "filtering," which, though also developed in *Principles of Chemical Engineering*, did not raise any major theoretical problems and in any case had little bearing on high-pressure chemistry.[63]

The engineers realized, however, that unit operations, even the most sophisticated ones, had their limits, for this concept essentially used a physics approach to chemistry, largely focusing on equipment and on the quantitative relations between incoming and outgoing fluids. Yet the notions of fluid dynamics, heat transfer, and diffusion can be gathered into a vast set of what were to be called "transport phenomena" after the war, but that were originally known as "unit processes."[64] These phenomena can be studied by calling on two major theoretical domains: the study of molecules, which makes it possible to describe viscosity, thermic conductivity, and other molecular notions; and thermodynamics and applied mathematics, which allow for an analysis of the velocity and the general behavior of fluids and gases by means of differential equations. Once the differential equations and the models of molecular flow have been established, the production operations can be carried out efficiently. This work of the chemical engineers in the Department of Chemical Research erased the traditional boundaries between applied and basic research by making use of basic studies to increase the efficiency of certain production processes. One of these chemical engineers, for instance, Allan Colburn—he had obtained his Ph.D. at the University of Wisconsin as a student of Olaf Hougen, a specialist in transport phenomena—was hired in 1929 to join Thomas Chilton's working group, where his interest in mathematics turned out to be useful in creating a model of the heat exchange in turbulent fluids.[65] This approach was related to certain unit operations, namely, hydrodynamics, heat transfer, and diffusion. But the novelty was that these were now grouped together and described by the same mathematical equations on the basis of the analogies among different unit operations. Transport phenomena thus provided a theoretical underpinning for the unit operations used to produce ammonia, nylon, gasoline, and eventually, plutonium.[66]

This theoretical approach to the problems of chemical production became important in the field of high-pressure chemistry, where the behavior of fluids and gases had to be analyzed as rigorously as possible. After catalysis, ammonia, for instance, presents itself as a mixture of liquid and gas at very high temperature, but its behavior was not well known. Similarly, the manufacture of polymers required the treatment of viscous and unstable liquids whose temperatures had to be precisely controlled. Here the traditional methods of control and adjustment proved to be too imprecise. The differential equations developed by Chilton and Colburn made it possible to solve this problem by establishing correlations among heat transfers, friction against the walls of tubes and vats, and pressure in the liquid.

In the space of about fifteen years (1920–35), these engineers achieved what they had set out to do. They made their profession a scientific discipline by linking it closely to chemistry, physics, and even mathematics, to such an extent that chemical engineering could no longer be practiced without a theoretical foundation. What had been an uncertain practice as late as 1920 had by the mid 1930s become a profession where scientific knowledge intersected with the engineer's empirical methods. Unit operations as defined by Little, Walker, and Lewis were profoundly transformed: some of them were pruned out as the dead wood of chemical engineering, while others were regrouped, deepened, and given a mathematical armature. Unit operations had replaced a catalogue of recipes with an orderly sequence of operations; now an orderly sequence of practices was replaced by differential equations that provided the means of solving varied and complex problems in an analytical or numerical manner.[67]

Yet chemical engineering's turn toward hard science was more than the mere effect of technological necessity. In the final analysis, this evolution resulted from a twofold movement. There is no question that, on the one hand, high-pressure chemistry required a new approach to chemical production. Differential equations are necessary for understanding heterogeneous solutions, and without thermodynamic calculations, it is impossible to produce ammonia or gasoline in large quantities. I do not mean to subordinate this technical evolution entirely to social configurations, yet it must be said that this new approach was promoted by the young chemical engineers, because it allowed them to find their place within DuPont—even if their claims to power were still confined to their professional practice. The turn toward scientific technology in chemical engineering was conceived and tirelessly pursued by a group of engineers who saw it as a way to strengthen their position at DuPont. Although they had the support of MIT professors of chemical engineering, with whom they were in regular contact, this endeavor was also something of a burden. As Greenewalt recognized, no one actually made him concentrate on high-pressure chemistry.[68] These men's desire to affirm their professional identity intersected with a technical intensification brought about by high-pressure chemistry.

In fact, and more generally, contrary to what is usually assumed, no technological or scientific evolution is natural: "progress" is the fruit of arbitration among professional or social groups, one modality among others of a social dynamic. Indeed, the very notion of progress should be shunned by the historian, to the extent that it tends to objectivize technical efficiency and to link technological evolution and the history of societies in one and the same progressive teleology. Rather than progress, one might use the notion of technical intensification, which recognizes certain lines of macroscopic force in the history of practices and objects, without, however, systematically overdetermining every action in the name of its assumed rationality in the march of progress.[69] In the same manner, technological determinism—which would consist of seeing new technologies as autonomous historical factors unrelated to social circumstances—is a sterile concept for the historian, even if it would be unwarranted to deny the existence of certain technical exigencies. What we really have to do is to avoid a position that sees either the implacable evolution of technology or specific but always reversible social configurations as the only trigger of historical change. Similar dynamics were at stake in other companies and in other specialties as well. One only has to think, for example, of American machine tools, which became increasingly sophisticated beginning in the 1950s (thanks to the introduction of numerical commands): according to David Noble, this evolution translated the will of a group of engineers to exert control over skilled and often organized and strike-prone workers as much as it did a technical need (in building supersonic jets) for a technical mutation in this direction.[70] In the case studied here, class was not obviously at stake—there are no archives with which to explore the working relations between engineers and chemical workers—but issues of power and professional assertiveness were not unknown.

Historians rightly consider DuPont a high-technology company, in truth one of the most advanced in interwar America. However, by varying the scale, one uncovers sharp contrasts among departments and brings to light different cultures coexisting within the same firm. Reinforced by its decentralized structure, these are visible as a series of strata superimposed on each other since the company's founding in 1802. It is therefore important for the historian not only to compare a company to others of its kind (in terms, for instance, of their technological advance or their management and organizational capabilities) but indeed to recognize heterogeneous units as vestiges of several layers of time within the company studied itself.

Hence the importance, in doing business history, of paying careful attention to the scale of the analysis. Enlarging the scale to the departmental level in DuPont's case reveals a gamut of practices ranging from the know-how of the shop floor inherited from nineteenth-century gunpowder manufacture to new forms of know-how based on formal scientific information, whereas Alfred Chandler's focus on the companywide scale flattens all heterogeneities and postulates an overly linear model of development for the organization. And yet the historian of organizations must ferret out disagreements, oppositions, jockeying for power within the structure, noise, and dissonances. A case in point is that in the early 1930s, the chemical engineers had by no means achieved their professional goals, because the depression affected sales, profits, investments, and research and development budgets. Bolton called every day, one of the engineers recalls with a sigh, bluntly asking what they had done to justify their jobs and their salaries.[71]

Under these circumstances the development of the first synthetic fiber gave the chemical engineers a chance to demonstrate their skill and the effectiveness of their approach, as well as the possibility of strengthening their ties to the chemists. The new culture of chemical engineering overtook the culture of the shop floor when it became associated with polymer chemistry; and eventually an accumulated technical capital—scattered and fragile though it still was—was in a position to crystallize around a great commercial venture.

Hand-Made: Nylon and the Mainstreaming of the Chemical Engineers

The invention and development of nylon, the first synthetic fiber, were the fruit of the work done by DuPont's chemists and chemical engineers between the early 1930s and 1938–39. The production of nylon involved the use of large numbers of chemical processes related to intermediary chemicals and fibers. These processes required heavy investments of capital and research, which finally paid off with the commercial triumph of nylon when it was marketed in the form of stockings starting in the spring of 1939.

The history of the research and development of nylon is well known, thanks to brochures, technical articles, and works of popular science published by DuPont immediately after the war, for the company considered nylon from the very beginning to be cause for celebration as its most extraordinary technical achievement.[72] Moreover, the premature and tragic death in 1937 of Wallace Carothers, the great chemist and inventor of nylon (he took his life after several years of depression) brought a memorable dimension to the story and further added to the legendary character of the synthetic fiber. In 1946, while preparing a brochure, DuPont's Public Relations Department actually asked the chemists who had invented nylon twelve years earlier to reproduce their experiments in the presence of a photographer, so that they could be preserved on film.[73] Here was a chance to give the epic of nylon a place in history and legend, to freeze an official account that presented it as a blessing of modernity.

Historians too have taken an interest in nylon, particularly David Hounshell and John K. Smith, who have published the most detailed technical account of its creation. They have shown how the success of nylon established a model, the "nylon model," under which DuPont subsequently endeavored to create "other nylons," that is to say, completely new products, rather than to improve existing ones. This strategy of radical innovation by means of heavy investments in basic research did not always pay off. Thus, after the war, DuPont launched synthetic fibers, such as delrin, that were spun into an inferior cotton substitute.[74] Commercial success is often achieved through patient and modest improvements to existing products rather than by spectacular novelties.

But the importance of nylon is not revealed in this perspective, which only deals with the management of the firm's research efforts. Jeffrey Meikle understood this when he wrote his study in cultural history describing the commercial importance of nylon and DuPont's marketing strategy in the first months after its launching in 1939, before the war temporarily interrupted its sales.[75] The Golden Gate International Exposition in San Francisco and the New York World's Fair in 1939 and 1940 provided opportunities to present the new fiber to the public and to stir up a feverish wave of early purchases. However, the first advertising campaign, which focused on the marvels science was bringing to humanity, was partly misunderstood. Consumers were upset by DuPont's claim that nylon was better than silk—in other words, better than nature. The firm therefore promptly changed its campaign and, instead of stressing the mysteries of polyamide chemistry, which seemed to be frightening ordinary people, extolled the practical and aesthetic qualities of nylon stockings. As Meikle writes, nylon had been domesticated, "transferred from the hands of the expert in molecules to those of the buyers."[76] The attention had shifted from the material (nylon) to the consumer item (the stockings).

This remark is important, for it suggests that the intrinsic qualities of nylon (and as we shall see, they were somewhat controversial) were not what assured its success, but that it was the marketing campaign orchestrated by DuPont in the favorable context of interrupted silk imports from Japan that persuaded initially skeptical consumers to buy. Nor was this all. Before it ever came to market, the way the new fiber crystallized DuPont's technical and human capital by stabilizing the disposition of men and knowledge within the firm also contributed to its success.

The Invention of Nylon

Historians of industrial research are aware of the importance of the decision of DuPont's Executive Committee to assemble several small teams of high-powered chemists in its Research Laboratory to do basic research.[77] This decision was made in response to a report written by Charles Stine, director of the Department of Chemical Research, and submitted to the Executive Committee on 31 March 1927.[78] Urging the company to explore certain areas of chemistry neglected by the universities, Stine advanced four major reasons why this should be done: the publication of scientific articles would bring recognition and prestige; this in turn would improve the morale of the company's chemists; results from this research might be useful as bargaining chips in agreements with other institutions; and, finally, practical applications might result from such research. No doubt in order to avoid alarming the management, Stine did not use the expression "basic research" in his presentation, but rather spoke of "pioneering research," which was considered more likely to lead to practical applications. To be sure, the distinction between "basic" and "applied" research is not really pertinent anyway, since the two are always closely intertwined. Nonetheless, Stine made it clear that all forms of science should be present in the organization.

Following the approval of the Executive Committee and the appropriation of the necessary funds, several small research teams were set up, each to work on a major area of chemistry. One worked on colloid chemistry; another was occupied with catalysts; a third focused on organic chemistry under the leadership of a young chemist, Wallace Carothers, who had studied for his doctorate at the University of Illinois as a student of Roger Adams and had subsequently been hired as an instructor by Harvard University. Carothers was then recruited by DuPont, where he went to work in February 1928 and stayed for nine years.[79] At Stine's suggestion, Carothers soon specialized in polymer chemistry.

A polymer is a giant molecule. Chemists had long believed that such molecules did not exist, and that what they were seeing were clusters of molecules loosely connected to one another according to the so-called micellar theory. But then, in the 1920s, the German chemist Hermann Staudinger showed that polymers do exist, namely, as macromolecules formed by complex chains of atoms (he was eventually rewarded with a Nobel Prize in 1953). By the end of the 1920s, most American chemists had rallied to Staudinger's thesis, which was supported by various studies based on crystallographical analysis or on the use of centrifuges, which made it possible to isolate enormous molecules, such as hemoglobin.[80]

Meanwhile, the industrial production of polymers was still part of an industrial history going back fifty years. The polymer industry was born around 1907, when the Belgian American chemist Leo Baekeland (1863–1944) invented Bakelite, a resin obtained by polymerizing an acid, phenol, by heating it in a kind of pressure cooker, the "bakeliser." Initially intended as a replacement for the ivory of billiard balls, Bakelite found a large market, in particular as insulation for electric wires and as material for the first radios.[81] Subsequently, different and internationally tested processes for manufacturing synthetic rubber grew out of Baekeland's pioneering efforts.[82] Until the advent of nylon, this polymer industry remained highly empirical, essentially working with the "recipes" of chemists without advanced knowledge of chemistry. Indeed, the very concept of the macromolecule was unknown before the pioneering work of Staudinger.

In the final analysis, then, one of the essential aspects of America's industry in the interwar period was precisely this: the rise of a kind of industrial research that now operated in the universe of the axiomatic language of the experimental sciences and of mathematics. This was what gave American industry new impetus after a quarter-century of continuing to live by the trial-and-error empiricism of its beginnings. In chemistry, the emblematic figure of this development was Carothers, the first high-powered chemist to be hired by DuPont, and the man who completely revamped the polymer industry by placing it on formal scientific foundations. As Bolton admiringly put it: "Carothers read from the depth of organic chemistry such as I have never seen. I think that he is the smartest organic chemist that the Du Pont Company ever had."[83]

Carothers's objective was purely scientific (feeling a certain contempt for applied research, he had accepted DuPont's job offer only after considerable hesitation): what he wanted to do was to create polymers in order to support Staudinger's thesis, while also shedding light on the nature of the bonds between their atoms. As it happened, too, IG Farben in Germany hired another young polymer expert, Hermann F. Mark, in 1927 to explore and make use of Staudinger's analyses. In 1938, Mark emigrated to the United States, where he became a recognized authority on polymer chemistry and also a consultant for DuPont.[84] Staudinger, Mark, and Carothers knew one another and were members of a chemistry "international" that had grown up around international congresses over the preceding fifty years.

In the spring of 1930, Carothers and his team invented two promising polymers. The first was neoprene, a synthetic rubber that was to prove crucial during the war, when imports of natural rubber were interrupted. Neoprene could then be used to make tires for Jeeps and bombers, soles for soldiers' boots, and many other essentials for the war effort.[85]

The second was formed by a very long molecular chain elaborated by Julian Hill, one of Carothers's assistants. A famous photograph shows Hill with a test tube, from which he is extracting a stringy, elastic whitish paste, which remained elastic but also became very strong after it had cooled. Hill later marveled at the memory of seeing the molecules lining up in parallel rows.[86]

From that moment on, Stine's objectives began to change. He asked Carothers and his team to stop working on the problems of polymerization, even though they had been explicitly hired to do just that, and instead to concentrate on selecting one chemical combination that would lend itself to industrial applications. This change of direction was confirmed in June 1931 by the appointment of Elmer Bolton as head of the Department of Chemical Research. Bolton was quite open about his disinterest in basic research in a chemical company. In his opinion, this type of research was the prerogative of the universities, and DuPont for its part should limit itself to developing industrial and commercial applications.[87] This rapid changeover from a research program of the academic type to one of research and development was actually not peculiar to DuPont. Historians of industrial research have already noted that programs of pure research rather quickly turned into applied research as soon as an attractive commercial perspective came into view. This has been shown, for instance, by Leonard Reich with respect to the research activities at General Electric and at Bell in the early years of the century.[88] After all, an industrial laboratory was not a "philanthropic asylum," to use the famous dictum of Eli Whitney, director of research at GE. One result of this renewed focus on commercial applications was that as early as 1932, half of the chemists hired by Stine had already left, and Carothers himself was seriously considering returning to Harvard as professor of organic chemistry, as he confidentially told Roger Adams.[89] Not happy with the turn toward commercial application forced by Bolton, the researchers were no longer sure of what was expected of them.

Between the invention of the fiber and its industrial development, there was a period of five years during which the researchers were undecided about which chemical combination to choose—or indeed turned toward other problems. Since the first combination (a polyester) had the flaw of melting at a moderately high temperature, the chemists finally decided to concentrate their efforts on the family of polyamides. In February 1935, they came up with a usable combination called "polymer 6-6," whose physical properties, such as elasticity and strength, proved very attractive, although it liquefied at a temperature of 263°C (505.4°F), which foreshadowed the need for complex manufacturing processes.[90]

Obviously, such moments of discouragement were not rare. And of course doubts and interruptions are not peculiar to the invention of nylon. Madeleine Akrich, Michel Callon, and Bruno Latour have pointed out that the stories of innovation that are told after the fact impose order on what at the time was sheer instability, a jumble of decisions and uncertainties concerning the very nature of the final product.[91] The official history of nylon, intent on presenting an edifying and prescriptible model of research and development, depicts an overly harmonious progression of scientific and technical steps. In doing so, it fails to show those of its aspects that are too tangled to fit into any model. This is too bad, for the hesitations, doubts, and compromises involved in a research project are often the reasons for its success, rather than shameful obstacles that must be hidden from view.

Nylon was the first major achievement to come out of DuPont's research activities. Unlike the electrical industries, American chemical companies had not particularly distinguished themselves until then. DuPont, for example, had had a great deal of difficulty diversifying in the field of dyes and had been obliged to go to great expense and great trouble to hire German experts, whom it transferred across the Atlantic in cloak-and-dagger operations right after World War I. So the success of this first attempt, the invention of nylon, and almost simultaneously that of the synthetic rubber neoprene, was a huge achievement for a company whose research activities were not comparable to those of IG Farben in Germany. This being the case, it is likely that DuPont was somewhat carried away by this success and that its directors came to feel that the company's expansion should be predicated on radically new products and massive investments in pure chemical research, all of them legitimated by a founding story, the exemplum of nylon.

Developing the New Fiber

In 1935, nylon was still a laboratory curiosity. The next step was to transfer it as rapidly as possible to the level of industrial production. This was easier said than done. It often happens that laboratory chemists invent a new fiber (or anything else) only to learn that the engineers consider the cost of producing it too high, or else that the marketing people do not think that it has any future. They then have to convince their professional partners that this novelty is indeed worthwhile and construct an industrial and commercial project likely to arouse their interest. This process of negotiation and project building usually does not take place in an orderly and linear manner. Rather than because of its intrinsic qualities, innovation comes about through the "art of enlisting a growing number of allies who will make you stronger and stronger."[92] A project is not necessarily brought down by enemies determined to scuttle it, but it does need active support. Failing that, it will grow feeble and disappear into some drawer. Innumerable promising projects are immediately set aside, not because of insurmountable technical obstacles or a lack of commercial prospects, but because they have failed to mobilize sufficient interest and to foster alliances that might have brought them to life. If nylon quickly became a commercial product, it was because its development mobilized a maximum number of DuPont employees who for one reason or another took an interest in it.

Whom, then, can we count among those who had decided to further the synthetic fiber project? To begin with, there were the company's directors. The Executive Committee kept informed about the work, authorized large expenditures, and generally pushed for results by questioning Bolton on a regular basis. Its strategy seems to have been effective: it set a course toward producing a synthetic fiber as rapidly as possible and then left further decisions to those in charge of the project. Funds were not unlimited, but it appears that Carothers's scientific prestige did a great deal to keep the financial spigot open, even in the Great Depression. But once the "6-6" combination was in place, its rapid development became a priority. The essential technical means were chosen by Bolton as director of Chemical Research, in conjunction with a steering committee consisting of the directors for this project (Bolton, Benger, his assistant Tanberg, the director of the Experimental Station and his two assistants, and the research supervisors, among them Greenewalt). This committee regularly invited the chemists to present their work—and this sometimes caused problems, often because of Bolton's autocratic personality—but eventually a dialogue between chemists and engineers did come about.[93] Exactly like nuclear physics some years later, polymer chemistry needed the support of chemical engineering to transform a scientific experiment into an industrial process.

The Ammonia Department was another key player. Recall the financial difficulties of this department, whose activities in the production of ammonia and its derivatives limped along despite the expertise of its engineers. As it happens, the intermediary chemicals needed for nylon (adipic acid and hexamethylene diamine) are obtained by means of catalytic reactions under high pressure—one from phenol, and the other from adipic acid and ammonia. They are then combined in a single solution, called "nylon salt." Manufacturing them was a problem, however, for adipic acid was produced only in Germany, and hexamethylene diamine was not available anywhere.[94] Several departments offered to manufacture these intermediary chemicals, among them the Department of Explosives. The Executive Committee had to arbitrate and assigned this production to the Ammonia Department, whose Belle plant was functioning at a very low level at that time.

This assignment was a godsend to the Ammonia Department, for it once again brought it profits and also restarted Roger Williams's career with DuPont.[95] The investments in high-pressure chemistry now paid off and proved useful for developing nylon. What the engineers of the Department of Chemical Research had been looking for in the 1920s—a product using catalytic processes under high pressure—had finally materialized. By the end of the 1930s and in the 1940s and 1950s, the manufacture of intermediary chemicals for nylon accounted for half of the sales volume of the Ammonia Department and about the same proportion of its profits.[96]

William Lazier of the Department of Chemical Research was in charge of developing the manufacturing process for hexamethylene diamine under the supervision of Greenewalt and in close contact with Williams.[97] Following a first study by Chilton's chemical engineers, some remodeling and some modifications of the existing equipment at the Belle plant—at a cost of nearly \$2.5 million—made it possible to manufacture the two chemicals in semi-industrial quantities. The full industrial stage was reached in late 1938.[98] DuPont also had to design special railroad cars to transport the nylon salt, which was dissolved in hot water so that it could be handled more easily, to the Seaford, Delaware, plant on the other side of the Appalachians, where the fiber was manufactured.

Nylon did not come to the ammonia specialists among the chemical engineers by chance, like some kind of divine surprise. It was because they had created a favorable technological terrain that it could so rapidly become an industrial reality and assume its full role as an item of mass consumption. There is an important connection between the technical and human capital accumulated during the interwar period and the expansion of the mass market for consumer goods, which could include products of very high added value. Note, by the way, that the connection between ammonia and nylon was intuitively made by advertisers and the public, for, just as they had done of synthetic ammonia right after its invention, people spoke of nylon as "made of air, coal, and water." Thanks to nylon, the Ammonia Department came out of the Great Depression in a position of strength.

The next step was to transform the nylon salt into a marketable fiber. This involved three stages: the polymerization operation in which polymers were obtained from the small molecules of the nylon salt; transforming the polymer paste into thread; and lastly, spinning the nylon.

Here two protagonists indisputably took the lead: the Department of Chemical Research and the Department of Rayon. The former was in charge of polymerizing the basic ingredients, as well as transforming them into thread, a complex task that required several months of intensive work by several dozen of this department's engineers and chemists. Their participation was to be expected, since it was in their department that Carothers and his colleagues had invented nylon. In the past, this department had been criticized for its relative inefficiency. Bolton therefore seized this opportunity to validate his department by making sure that it would be involved—directly or indirectly—in every stage of the research and development for this project. "The great time was in the nylon days!" Greenewalt later exclaimed.[99] By this he meant to convey that nylon, along with neoprene, represented his department's first great success, and that he could not pass up the chance to show it off.

Polymerization is achieved by means of an autoclave, in which very high temperature and pressure bring about the bonding of macromolecules. The experimental autoclave and later the production autoclaves built by the DuPont engineers were still quite similar to Baekeland's bakelisers. However, it was necessary to keep exact control of the size of the molecules and of the viscosity of the preparation, and this initially caused some problems when the polymer obtained was of too high a molecular weight. Then someone had the idea of stopping the polymerization at the right moment by means of an acid.

When this was done, the chemists finally obtained a thick and translucent paste of the consistency of honey. The next step was to transform it into a fiber. To that end, this polymer paste first had to be liquefied, either by adding solvents or by heating. The first option was fairly rapidly discarded because solvents such as phenol would have been difficult to recover later, after the solution had solidified. The other possibility was to heat the polymer, but its very high melting point—about 285°C, or 545°F—caused the formation of gas bubbles that would break the burning hot filament that was expelled through the tiny holes of the spinnerette (a steel plate pierced with holes through which thin streams of liquid ran out) and solidified immediately upon contact with the air. Moreover, the holes became clogged with the cooled solution. The answer, worked out by George Graves, was to keep the polymer liquid under high pressure, which prevented the bubbles from forming and thus yielded a continuous stream. This liquid was then passed through the spinnerette, and when the thread that was obtained had cooled, it was slowly drawn out and—a marvel of polymer chemistry!—retained its new dimensions yet remained elastic and solid. This was the "cold drawing" method.[100] In May 1937, DuPont's engineers were in a position to produce a continuous nylon thread.

One would expect the Department of Rayon to look askance at the entry into the market of a potential competitor to rayon, which was still going strong. And indeed, the reticence of the rayon men is perceptible. It showed itself, for example, when one of them, Preston Hoff, paid a visit to the nylon laboratory. Hoff reported to his superiors that "the rate at which its problems are solved" was not commensurate with the hopes vested in nylon; he was also skeptical of the commercial prospects of the new fiber.[101] But these fears were soon put to rest by the decision to position nylon as a luxury product that was not only considerably more expensive than rayon, and even silk, but in addition was projected to be made into stockings and lingerie, for which rayon was not suited. Under these circumstances, the Department of Rayon could cooperate fully in the development of nylon. It wove the first pieces of material, manufactured the first stockings, and corrected the flaws pointed out by the first female testers, usually DuPont employees or wives of employees.

The last ally was the Department of Engineering. Benefiting from its solid experience in construction, this department distinguished DuPont from other American chemical companies, which entrusted the building of their plants to public works firms backed up by consulting engineers. This department's special qualification, incidentally, was not lost on General Leslie Groves, head of the Manhattan Project during the war; and Groves was an expert such matters, being an officer in the Army Corps of Engineers who in 1941–42 had overseen the building of the largest low-rise office complex in the world, the Pentagon.[102] DuPont's Engineering Department had been headed since 1927 by Everett Ackart. Born in 1881 near New York, Ackart earned a degree in mechanical engineering from Cornell in 1902 and worked on the design and maintenance of electric generating stations for New York City for several years before going to work for DuPont in 1907 as an engineer in its Heating, Lighting, and Electric Current Division. Subsequently, he worked in the Engineering Department, where he rose through the ranks, especially after he was given the essential responsibility for the design of the plants, and he eventually became the chief engineer for these projects.[103] Ackart was a typical representative of a generation of mechanical engineers hired by DuPont before it was felt that large numbers of chemical engineers were needed. In the early 1920s, many of DuPont's mechanical engineers had left to go to work for General Motors, although some had stayed and new ones were hired, particularly for the Engineering Department.

For the Engineering Department, building the pilot plant and the commercial factory was an opportunity to work with new processes with which it was not familiar. Competition between the engineers of the industrial departments and those of the Department of Engineering could be intense, for the latter wanted to become involved wherever they could, especially in new projects considered to be of prime importance by the Board of Directors. After the war, this interference by Engineering was the subject of several reports that detailed conflicts of authority and a perception of the engineers' arrogance.[104] But for the moment there was no shortage of money or men to mass-produce nylon: about 1,300 tons were produced in 1940.[105] The building of these plants required closer collaboration than usual between the Departments of Engineering and Chemical Research. A special building program was authorized. As a first step, a pilot plant was erected close to the Experimental Station. A pilot plant was nothing new for DuPont—the firm had built one as early as 1907–9, when it was readying its first production unit for TNT—but this was not a widespread practice.[106] Next came the commercial plant at Seaford, completed on 27 October 1938, which was designed and presented as the very model of a modern chemical plant. The 1,800 employees enjoyed the use of a cafeteria and air conditioning. Nylon was meant to embody modernity even at its manufacturing site.

The success of the nylon project thus had to do with its ability to achieve the rapid mobilization of a large number of DuPont's chemists and engineers, all of whom were interested in furthering it for one reason or another. Because of the interdepartmental collaboration required by the complex development of the synthetic fiber—DuPont registered some 500 patents—the roughly 230 chemists and engineers who were directly involved had a chance to prove themselves, get to know one another, and coordinate their work.

After some hesitation and at Bolton's instigation, DuPont's directors decided to focus on the production of nylon for the manufacture of stockings, as well as, secondarily, for toothbrushes. This marketing decision was not fortuitous. Stockings had traditionally been made of silk imported from Japan, and in 1938, more than 700 million pairs were sold. But by the late 1930s, Japan was seen by the Americans as an aggressive and expansionist power that some day might well come into direct conflict with the United States. Bringing nylon stockings onto the market was thus a shrewd move. However, the sales pitch could not be based on price, since nylons cost practically twice as much as silk stockings, in keeping with the cost of their raw materials: $4.27 per pound of nylon against $2.79 per pound of silk. Hence the new fiber had to be presented as more agreeable to wear and more practical than silk, for it could not be called stronger either: nylon stockings were—and still are!—fragile, not because the thread is not strong but because of their tendency to "run," or unravel lengthwise. The fact is that nylon stockings were not really better than silk stockings, and the craze they stirred up was primarily due to a carefully orchestrated publicity campaign that vaunted the qualities of this "artificial silk" in a context of defiance against Japanese silk.

On 24 October 1939, the day when the first nylon stockings were sold to the public in a store in Wilmington, Delaware, some women shoppers came all the way from Philadelphia or even New York. The first nationwide sale took place on 15 May 1940, a memorable day, which the *Du Pont Magazine,* in its full-throated enthusiasm, later suggested calling "N-Day." Over the course of the next year, some 64 million pairs of nylon stockings were snatched up. From the very beginning, nylon was an enormous commercial success, which further enhanced the success of the research and development effort that had gone into it.

The technical and commercial success of nylon, which became clear as soon as it was brought on the market in May 1940, six months after the Golden Gate International Exposition in San Francisco and during the New York World's Fair, reinforced the DuPont directors' conviction that radical technical innovation was the key to the company's development. That success also made chemistry into a "star technology," on a par with electricity, and became a launching pad for those who had worked on nylon.[107]

Thus the development of nylon, even more than its invention in the laboratory, bears the stamp of a remarkable achievement, combining as it does an innovative product with innovative processes, which was the highest goal of industrial research in the prewar period. Like the tungsten filament light bulb brought out by General Electric in the 1920s, nylon gave the public a high-tech product of mass consumption. It is not a coincidence that of the first two synthetic polymers that DuPont had put on the market, nylon and neoprene, only the former captured the attention of the crowds. To be sure, they were announced to the public in very different ways: nylon was unveiled at a preparatory meeting for the New York World's Fair, while neoprene was announced on the occasion of a meeting of the Rubber Division of the American Chemical Society.[108] While the press did report this announcement, there was no comparison to the high excitement caused by nylon and the huge advertising campaign with which DuPont proceeded to launch it. All of this greatly impressed the female customers, who were seduced by the good looks of the stockings and their reputation for being indestructible and mysteriously made out of "coal, water, and air," as the ads boasted.[109]

Nylon certainly was not the first artificial fiber, since rayon had preceded it. But rayon was made from cellulose, that is to say, from wood pulp, and thus came directly out of the prewar strategy of diversifying on the basis of the original know-how for nitrocellulose. Rayon was actually a distant cousin of dynamite. The diversification that came out of cellulose had already moved DuPont away from its original vocation—which it had been called upon to pursue by Thomas Jefferson himself. But in a sense this diversification was a natural continuation predicated on synergies of different products and on improving products and processes purchased abroad, usually from France or Germany. But then nylon marked a complete break: it had nothing to do with cellulose and nitric acid and was grounded in different categories of scientific and technological knowledge. One of these was polymer chemistry, a field of which DuPont, along with most other American chemical firms, knew very little before it hired Carothers in 1928; and the other was high-pressure analytical chemistry, which, introduced by Williams and his Ammonia Department, had been fraught with the difficulties outlined above. Nylon crystallized a hitherto dispersed and fragile technological capital that had been amassed during the interwar period around one end product; and this development validated a whole gamut of heterogeneous bodies of knowledge. In the spring of 1940, the fibrous weft of the nylon stockings that so delighted American women seemed to symbolize the fine meshwork that had woven together a company, a group of professionals, and a body of technical knowledge.

A direct consequence of this break brought about by nylon was the definitive marginalization of the culture of the shop floor that had once been such a distinguishing feature of the American chemical industry. To be sure, until the period after World War II, the Departments of Gunpowder and Explosives preserved their own culture, which was founded on the workers' experience and know-how. But these departments no longer counted for much at DuPont, although they remained fairly profitable financially. One can see this in the *Du Pont Magazine,* a monthly publication that featured the company's marvelous products. Until the early 1920s, articles about dynamite and gunpowder for hunting weapons were largely predominant. Explosions were shown on practically every page. After the end of the 1920s, and particularly in the 1930s, explosives yielded to items of mass consumption, such as plastics, cellophane, and antifreeze.

Du Pont's young chemical engineers had concretely demonstrated to the skeptics that they were not just theoreticians for whom there was no practical use, but rather employees indispensable to the company's commercial growth. DuPont's directors, taking note of the achievements of their chemical engineers, asked the universities to create specific training programs dealing either with research problems, problems of production, or perhaps even administrative issues, with a view to turning chemical engineers into future managers. In a characteristic speech delivered on 11 November 1938, Charles Stine considered this possibility:

> We wonder whether we might not expect to obtain much better results by differentiating among the better types of schools offering a course for chemical engineering training, finding perhaps on the one hand a school whose objective it is to develop research chemical engineers, perhaps a second type of school with the objective of developing primary engineers well equipped for work along the lines of design and production, and perhaps a third type of school which has in mind not specifically the equipping of any of its graduates for specialized professional work in engineering, but desires instead to use engineering courses, and particularly chemical engineering, as a method of attaining a broad acquaintance with the problems of modern business practice.

. . . Thus the course for research chemical engineers would be strong in advanced mathematics, organic chemistry, and experimental methods; the course for design and production would be strong in machine design, electrical engineering, power generation and transmission, properties of materials, and safety engineering; the course for administrative engineers would include an extended course in the economic problems of chemical engineering, the history of the chemical industry, and a study of industrial and public relations.[110]

The mainstreaming of chemical engineering concretely showed itself not only in the hiring of chemical engineers but also in the early recognition of their future function as managers, and thus of their importance in shaping the company's general strategy. In concluding his speech, Stine did not hesitate, in fact, to assert that "to assume tomorrow's leadership, we choose the chemical engineer!"[111] This may have been too facile a way of playing to the gallery, since Stine was speaking to an audience of chemical engineers, who probably would have been satisfied with less. But there was something else: Stine, a DuPont director, felt that in twenty short years, the hiring of engineers trained according to the new principles, as well as the work they were doing for the firm, had demonstrated their importance to the company.

This institutional recognition extended beyond DuPont. Significantly, in 1939, Thomas Chilton received the Chandler Medal, a prestigious prize in chemistry that had notably been awarded to Baekeland in 1914, Whitney in 1920, Langmuir in 1929, and Conant 1931. Chilton was officially rewarded for his "remarkable merits in the discovery and the formulation of the principles underlying unit operations in chemical engineering, and for the application of these principles to the development and the design of chemical plants."[112] To this, the day's honoree responded with a speech that was a blend of technical considerations and flights of lyricism:

> The commercialization of these synthetic materials, from which the public benefits, is not miraculous but is the result of painstaking application of sound engineering, as well as of brilliant chemical research. These triumphs are no more miraculous, and no less so, than the George Washington Bridge, the Holland Tunnel, and the Empire State Building; the Zephyr trains, the Queen Mary, and the Atlantic Clipper; or the Hell Gate power station, the New York telephone exchange, and Station WJZ. The work of the civil engineer may thus be prominent, and his place, developed over a long period of years, is readily acknowledged. The mechanical engineer emerged after the invention of the steam engine and

the consequent transition from home industry to the factory system a hundred years ago. The metallurgical engineer found his calling when the blast furnace and steel mill supplanted the iron puddler and the smith. The electrical engineer filled the need resulting from the discoveries of Faraday, Henry, and Edison. Likewise researches in theoretical and organic chemistry, each paving the way for others, have demanded for their adaptation to practical use what we now call the "chemical engineer."[113]

Continuing his peroration, Chilton placed particular emphasis on "transport phenomena," citing several articles on the subject published by himself and Colburn, as well as studies by Edwin Gilliland and William McAdams at MIT: his classic presentation began with the ammonia synthesis and ended with the problems of turbulent fluids and mass transfer. Chilton, along with his colleagues in chemical engineering, had understood—and how would he not?—that his professional work had changed the chemical industry and that he deserved to be celebrated in terms that would have seemed ridiculous twenty years earlier. Henceforth, the chemical engineer could indeed be compared to his prestigious colleagues, the electrical and the civil engineer, who mastered the magic of electricity and the law of gravity.

The profession had come a long way from the uncertainties of the early years of the century. On the strength of this technical and human capital that it had built up over twenty years, DuPont had achieved its mutation from a gunpowder and explosives firm during the Great War to a company that had amassed enough technical knowledge to enable it to venture into new projects on the eve of a new war. We must therefore look upon this company as a site where knowledge was reordered: where knowledge hitherto confined to the academic world emerged at the scene of industrial production, there to be reinforced, redefined, and finally catalyzed by the development of a product that was barely born when it became a legend.

CHAPTER THREE

Culture and Politics at DuPont before World War II

At the end of the 1930s, DuPont no longer had much in common with what it had been at the beginning of the century, when the du Pont cousins took over. This event had brought about a technical and organizational mutation that changed the company from a simple producer of gunpowder and explosives into an enterprise focused on the high-volume consumer market for chemicals and a pioneer in advanced management techniques.

This mutation went hand in hand with a change in customers. The U.S. Army and Navy had been DuPont's principal customers until World War I, but when the company went into the high-volume civilian consumer market, production for the military declined proportionally. This orientation was in part brought about by the change in DuPont's relationship with the federal government, a change that went back to the very first years of the new century, when the Progressives were fashioning a new government that had greater powers of intervention and regulation than in the past. Until World War II, DuPont's relations with the government were fraught with tension and punctuated by periods of crisis that became acute in the first years of the new century, during World War I, and particularly in the early 1930s, the first years of the New Deal.

Moreover, DuPont had to deal with the public's moral disapproval. Articles in major daily newspapers stigmatized the "gunpowder trust" and the "American Krupps," and these attacks became even more prevalent after World War I and in the 1930s. On the eve of World War II, and just before the United States definitively assumed a leadership role in international affairs, DuPont was therefore politically on the defensive and extremely reluctant to venture into new projects with the public authorities, even if the lines of communication had been reopened. This was because the company's leaders had early on recognized the importance of developing a high-volume market for chemicals, while the New Deal for its part had adopted a Keynsian perspective, in which consumer spending became a political priority and attacks on big capital were relinquished. The "consumer" as a political concept was thus a highly visible figure when nylon came on the market; and this chronological concurrence, which created a symbolic seal between a political economy and an already famous product, prepared the way for the postwar era.

The historians who have taken an interest in the political activities of the du Pont family have concentrated on the American Liberty League, an anti-New Deal association launched with a great deal of fanfare by the du Ponts in 1934. They have rightly stressed the truly reactionary nature of this group and its lack of popular support.[1] Nonetheless, the political battles of the eponymous directors have not yet been integrated into a larger perspective that would go beyond moral considerations about the egoism of millionaires and allow us to establish the connection between the profound conservatism of these businessmen and the savvy boldness of their investments in high-volume consumer chemicals.

The du Ponts and the Progressive State

DuPont and Federal Regulation

As we know, DuPont at the time of its founding had very close ties to the American government, particularly the military. This close association continued throughout the nineteenth century, even if DuPont endeavored from the beginning to expand its activities into the civilian market for black gunpowder and explosives. But until the early years of the twentieth century, there were no plans to produce anything else but black powder, smokeless powder, and various kinds of dynamite. In 1903, DuPont sold 70 percent of all the dynamite sold in the United States, slightly more than 60 percent of the black powder, 65 percent of the nitrocellulose powders, 80 percent of the cannon powder, 70 percent of the powder for hunting rifles, and 70 percent of the smokeless powder (which was used almost exclusively by the military).[2] Its profits came essentially from the sale of dynamite and smokeless powder for the Army and the Navy. In 1905, for instance, of an overall profit of $4.88 million, $1.90 million came from dynamite sales and $1.07 million from military smokeless powder, with the rest evenly divided among various kinds of powders for rifles. In January 1907, J. Hamory Haskell, the manager of military sales, estimated that a profit of $3 million per year, or 60 percent of the total profit, was made on the sale of gunpowder to the Army and the Navy.[3] In commercial terms, DuPont's ties to the government were closer than any other major American company's.

The first break occurred in the early twentieth century, when the company's relations with the federal government rather suddenly deteriorated.[4] At that time, many Americans felt that the rising power of big monopolistic businesses, led by sometimes unscrupulous businessmen, was a threat to their civil liberties, to local communities, and in a general way to the democratic principles on which the nation had been founded. In terms of politics, the Progressive movement expressed itself in the election of politicians who favored substantial social reforms, to be implemented notably by increasing the regulatory functions of the government.[5] The Progressives felt that the federal administration had to grow if it were to deal effectively with socioeconomic problems that the "invisible hand" of the market could no longer control. Herbert Croly wrote in 1909 that the country must go from "natural abundance" to "organized abundance." It was a theme that was sounded again, mutatis mutandis, a half-century later by the historian David Potter.[6]

This new political approach assumed concrete form in the creation of a series of oversight and regulatory commissions for various industries. The first of these was the Interstate Commerce Commission, established in 1887 to interdict price agreements between railroad companies, and it was followed by other commissions set up to regulate the market for electricity, gas, and other commodities considered essential and liable to create monopoly situations.[7] The Progressives stressed the opposition between "the trusts" and "the people," and the Sherman Act, passed in 1890, stipulated in rather loose terms that "every contract, combination in the form of a trust or otherwise . . . is hereby declared illegal,"[8] which left it to the courts to provide the precise definition of what a "trust" was. One must not think that the trend toward regulation was uniform and systematic. The suits brought against Standard Oil and American Tobacco resulted in 1911 in the dismantling of these two companies, but many others, such as U.S. Steel and General Electric, managed to escape unscathed.

It was in this context that trouble first started for DuPont. Through a gunpowder cartel formed in 1872, the Gunpowder Trade Association, DuPont effectively controlled two-thirds of the powder and explosives production in the United States, especially since this cartel had subsequently been consolidated in order to form the DuPont de Nemours Powder Company in 1903.[9] Technically, DuPont was no longer a trust. The company was no different from large numbers of other firms that had been restructured at this particular time in order to rationalize their production and get around the Sherman Act.[10] This "Powder Trust," as the contemporary press called it, was under legal investigation from 1907 on, following the revelations of a former DuPont employee and articles by muckraking reporters who inveighed against giant companies and "robber barons." Note that, strictly speaking, it was no longer a question of a trust, but rather of a big company allegedly seeking to monopolize a trade. Thus it was a matter of finding out whether or not DuPont's quasi-monopoly had been built up legally.[11] An official investigation of possible violations of the Sherman Act in sales of smokeless powder to the military was launched on 31 July 1907. Congress took up the case and directed the Navy to build its own gunpowder plants, but DuPont's lobbying efforts in Washington, conducted by a Colonel Buckner, paid off, and no concrete steps were taken. But the trial was a different matter.

What happened was that after a lengthy investigative phase, three federal judges decided in 1911 that DuPont had to shed a certain number of its production units, which the following year were regrouped into two independent firms, Atlas Powder and Hercules Powder. Named after two varieties of dynamite, these firms recovered 42 percent of the dynamite market and 50 percent of the market for classic gunpowder. The separation did not, incidentally, prevent the du Pont family from controlling these two firms indirectly, but they nonetheless perceived the creation of the Federal Trade Commission in 1914 as one more manifestation of the government's hostility toward big business.

DuPont's reaction was inept. Instead of working for a compromise with the executive and legislative powers, its eponymous directors engaged in noisy maneuvers in the corridors of the White House, the Congress, and the Republican Party. The du Ponts conceived of politics as founded above all on personal relations and on having reliable people in Congress, where they would be able to block legislation unfavorable to the firm. T. Coleman du Pont activated his networks in Congress and at the White House, but to no avail. The fact is that a luncheon at the Metropolitan Club was simply no longer enough to influence decisions. Such a vision of politics was not suited to the new era, which required that companies replace the personal lobbying that had characterized entrepreneurial firms with a new, better organized, and more subtle style of lobbying. In February 1906, *Cosmopolitan* published "The Treason of the Senate," a series of sensational articles in which the author, the muckraking journalist David Graham Philips, excoriated this "millionaires' club" that safeguarded the interests of the trusts.[12]

One of Eleuthère Irénée's grandsons, Henry A. du Pont, known as "Colonel Henry" (he had been a lieutenant colonel in the Union Army during the Civil War), a partner in the firm before it was bought by the du Pont cousins in 1902, had been elected senator from Delaware in 1906. A member of the most conservative wing of the Republican Party, he had fought Progressive legislation, particularly the passing of labor laws, and supported a protectionist policy and high military budgets. He was defeated in 1916 by his Democratic opponent, but the family tradition was carried on by his cousin T. Coleman du Pont, one of Pierre's associates, who was appointed to this Senate seat in 1921, following some controversial political maneuvers.[13] In the corridors of the Capitol as in the halls of Wall Street, the name du Pont was often spoken, respected, and feared.

But the lobbying efforts of Henry, Coleman, and their men in Washington were founded on an older conception of political life. The twofold trend toward a more bureaucratic and more democratic conduct of American politics required a different practice, one that had already been adopted by companies like U.S. Steel and International Harvester, which under the direction of George Perkins and Cyrus McCormick, respectively, had shown their account books and their business dealings to federal agents of the Bureau of Corporations. Created in 1903, this agency was charged with investigating economic activities and making recommendations to the president. The heads of these companies had made concessions to the representatives of the government in exchange for protection from the bureau against action by the Department of Justice under the Sherman Act.[14] By taking the initiative, they had forestalled a possible antitrust action, and their firms had emerged unharmed from a critical period. DuPont, by contrast, did its best to complicate its relations with the government. As early as 1908, its directors required every representative of the government who visited any DuPont plant to swear that he would not reveal anything of the production methods he had seen. And when war broke out in Europe in 1914, DuPont refused to furnish a list of its overseas clients to the War Department.

In the same vein, the company, even before the war, launched a diversification into other areas than gunpowder and explosives in order to loosen its ties to its traditional customer as much as possible, for its directors no longer trusted the government and considered it hostile to free enterprise. Nor was this the only reason for diversification: DuPont also wanted to benefit from the emergence of a civilian mass market. On 16 December 1908, a meeting of the Executive Committee decided to form a committee charged with examining possible alternative uses for nitrated cotton and nitrocellulose, such as celluloid, simulated leather, and "artificial silk," the future rayon.[15] Fraught with important consequences, this decision coincided with a substantial decline in orders from the Army.

The Paradox of World War I

Paradoxically, World War I reinforced DuPont's strategy of diversification. And yet the main concern of its directors in the fall of 1914 was how to fill the huge orders placed by the Allies. Sales and profits rose tenfold during the war. The production capacity for smokeless powder rose from 350 to 19,000 tons per month between 1914 and 1918. New plants had to be built in record time, thousands of employees were hired, and huge orders for nitrate, sulfur, cotton, glycerin, and various acids were placed. Profits for the year 1916 alone were larger than those recorded between 1902 and 1914. During that period, profits had varied between $3.9 (1907) and $6.9 (1912) million annually. They rose to $57.8 million in 1915, then $82.1 in 1916, and $49.3 and $43.1 million in the next two years. By 1919, after taxes and dividend payments to its shareholders, the company had thus cleared $71.7 million.[16] The price of a share of DuPont stock rose from $160 in July 1914 to $319 in April 1915, $670 in June, $775 in September, at which time the shares split 2 for 1, and the new shares soon passed the $400 mark. World War I undoubtedly brought extraordinary and unprecedented profits to DuPont, as the company's directors were the first to acknowledge in congressional hearings after the war. This might have strengthened the position of those who wanted the company to continue most of its activities in the gunpowder and explosives sector.

But two factors made diversification an urgent concern. One was the directors' fear that these enormous production capacities would become useless
the day hostilities in Europe came to an end. Hence a series of studies was undertaken by the Planning Department in order to determine which products
could make use of the equipment for the war effort, and the celluloid manufacturer Arlington was acquired in 1915. A second point was, once again, that relations with the American government were less than smooth during the war.
A few months after the United States joined the war against Germany, on 30
July 1917, DuPont signed a contract under which it was to supply 60,000 tons
of smokeless powder to the U.S. government. By the fall of that year, the needs
of the military rose even higher, and it became necessary to build new plants
that would be owned by the government but managed by Du Pont. The negotiations soon came to grief over the percentage of the profits DuPont would
derive from the operation: the government felt that the 15 percent DuPont
demanded was too high a margin in view of the profits it had already pocketed
from its gunpowder sales to the Allies. But Secretary of War Newton Baker did
not really have a choice, since DuPont was by far the most qualified company
to undertake this gigantic production. Time was of the essence; the doughboys
were already landing in Europe. A contract was finally signed in January 1918
for the building and the operation of a plant dubbed Old Hickory, located not
far from Nashville, Tennessee.[17] The plant, by far the largest production unit for
explosives in the world, began to produce its first tons of gunpowder in June
1918 and was nearing completion when the Armistice was signed. By that date,
it had already manufactured 17,000 tons of powder.

Then, in 1921, an investigating committee of the House of Representatives
claimed that Old Hickory had not been necessary and had been "an enormous
waste and an extravagant outlay of public money. . . . Those responsible for it
deserve the disapproval and the condemnation of the nation."[18] Fingers were
pointed at DuPont for having made undue profits, and the company was accused of having lobbied for the building of Old Hickory—a charge against
which its directors protested vigorously. Even though World War I had brought
DuPont considerable profits, it reinforced its leaders' fears: the government
was decidedly a difficult and ungrateful client, and dealing with it was becoming increasingly troublesome. Moreover, these dealings exposed the company
to attacks in Congress, which were spread about by the press and harmful to
its business.

And to top it all off, tax disputes further aggravated matters. In 1916, Congress had passed a law, soon to be dubbed the "Munitions Tax," stipulating a retroactive 12.5 percent surcharge on profits from the sale of explosives and cannon powder, much to the distress of the du Pont cousins, who protested vigorously but unsuccessfully.[19] One daily newspaper mentioned—and this is confirmed by several authors, among them Alfred D. Chandler Jr.—that in the fall of 1916, DuPont actually forbade its employees to wear campaign buttons favoring President Woodrow Wilson's reelection.[20] The company denied this but at the same time asserted that the Wilson administration had been "unfriendly" toward it.[21] Another problem arose three years later in connection with the manner in which DuPont and other chemical companies that had banded together in the Chemical Foundation had taken possession of the German patents for dyes. The action by the Justice Department misfired, but it further increased the company's mistrust of the federal authorities, even though these were particularly well-disposed toward big business, this being the time of the Harding administration, in which the du Pont cousins had several close friends. They had also given financial support to Warren Harding's presidential campaign of 1920.

All in all, then, the war reinforced DuPont's civilian orientation. By acquiring other chemical firms and making a strategic investment in General Motors, it was able to point the investment of its remarkable profits toward a civilian market that seemed more promising and more secure in every respect than the exclusive preserve of supplying gunpowder and explosives to the military. A number of considerations concerning the growing attractiveness of the civilian market and the troubles encountered in the military market came together to support this major displacement from one market to another.

Even beyond their political and legal skirmishes, the du Ponts were totally at odds with the federal government regarding the expansion of the government's administrative competencies.[22] Nor were they the only ones to hold this view, for mistrust of the federal government was a widely shared attitude, at least until World War II. The du Ponts had denounced the creation of the Interstate Commerce Commission, and then that of the Federal Trade Commission, as an unacceptable intrusion into the country's economic life. The anti-trust actions of the federal authorities had aroused their ire, as had the fiscal measures that had affected them during the war. Pierre du Pont had seen it as another proof of a dangerous expansion of the government that some day might turn into collectivism. But the du Ponts were not opposed to regulation, as long as it was organized by the industrialists themselves.

In the early 1920s, as the Progressive dynamic drew to a close and as a majority of Americans seemed satisfied with their prosperity, the model of a political economy favored by the leaders of industry and by Republican administrations was one in which business was regulated by trade associations, that is to say, by associations of companies in the same field that themselves set the rules to be followed. In this perspective, the public powers had no more than a modest role as sources of information that might bring up the matter of anti-trust laws from time to time.

But not everyone in the industrial community had exactly the same ideas about the role of the trade associations. Commerce Secretary Herbert Hoover, the former and very charismatic president of the Association of Mechanical Engineers, felt that the largest companies should not be allowed to take over absolute control of the market: the trade associations, he said, should not resuscitate the cartels of the late nineteenth century, the very reason why anti-trust legislation had been passed. Other personalities in the business community, notably Pierre du Pont, wished for a complete centralization of the associations and the creation of a "super-council" that would be similar to the War Industries Board of the Great War, but exclusively run by directors of major companies. In this respect, the special conference committee created in 1921, which brought together representatives of the ten (Bethlehem Steel, DuPont, General Electric, General Motors, Goodyear, International Harvester, Irving Trust, Standard Oil of New Jersey, U.S. Rubber, and Westinghouse) and later the twelve largest companies, and whose object was to combat "revolutionary syndicalism," must have been seen by Pierre du Pont as the very model of future cooperation.

Two major categories of political action by the du Ponts can be distinguished. First, there were family members who had become actively involved in politics, with two of them actually being elected senators for Delaware. T. Coleman du Pont in particular was a well-known figure in the political life of the early twentieth century; for a time, he even served as chairman of the Republican Party, and everyone was aware of his influence and his networks. Then there was Pierre S. du Pont, who possibly for psychological reasons—he was extremely shy and self-conscious because of a limp—and in any case for strategic ones was politically retiring: he never ran for office and expressed his political views discreetly, usually in his personal correspondence, for instance, in writing to his old friend John Raskob, the inspired financier for DuPont and General Motors. In the 1930s, he also spoke up at meetings of the various boards on which he sat. And finally, his two younger brothers Lammot and Iré-née, who succeeded him as directors of the firm, became more openly engaged in the 1920s and 1930s by working with organizations, such as the Association Against the Prohibition Amendment (AAPA) and the Liberty League, without, however, seeking elective office. The du Pont brothers operated at the margins rather than at the center of the public sphere and also disengaged from their military clientele at a time when the civilian market seemed to be headed for limitless growth.

The New Deal and Nylon

From Prohibition to the First New Deal

The first years of the New Deal were a period of particularly tense relations between DuPont and the federal government, and things only began to improve in 1938. The du Pont brothers frontally attacked Roosevelt, and their highly publicized broadsides had consequences for their company and further reinforced its identity.

Since the beginning of the century, the du Ponts had been generous donors to the Republican Party, especially during presidential campaigns. But in the course of the 1920s, a certain number of disagreements arose between them and that party. The most visible of these concerned Prohibition, which had been instituted at the national level in 1929 by the Eighteenth Amendment. This issue increasingly alienated the du Ponts from the Republicans, the vast majority of whom favored the temperance movement.[23] Beginning in 1920, the du Ponts and Raskob launched a full-scale anti-Prohibition crusade by taking the reins of the AAPA, which had been founded a few years earlier by the Democratic Party politician Jouett Shouse and William Stayton, a former Navy officer.[24] The motivation here was not so much a particular fondness for alcohol as the fear of a limitless expansion of the government's power. Moreover, the AAPA also documented the ineptness of Prohibition by raising a number of fiscal questions about the loss of income for the government and the cost of customs controls and police actions. Pierre du Pont favored the controlled sales of alcohol by a private monopoly.[25] The issue of Prohibition caused Raskob to support Al Smith, the Democratic presidential candidate and champion of the anti-prohibitionist cause in the presidential election of 1928—and even to become chairman of the Democratic Party. The du Ponts were divided on this issue, for Pierre supported Smith, while his two brothers remained faithful to the Republicans and to Herbert Hoover.

But one should probably not see Prohibition as the major issue of the moment, as the analysis of the historian Robert Burk would have it. Another and more surreptitious battle pitted two groups of companies against each other, as we learn from the analyses of the political economist Thomas Ferguson.[26] On one side were firms whose high concentration of capital made them perfectly competitive in the world market, firms like General Electric or Standard Oil, which militated for a lowering of import duties—free trade was to their advantage and they were not afraid of foreign competition—and were relatively accommodating toward labor and labor unions. On the other side were firms dealing with large labor forces and/or intense competition from foreign companies, such as the automobile and textile industries, which wanted to keep the customs barriers high. The chemical industry tended toward the second camp because of the preeminence of giant European companies like Imperial Chemical Industries (ICI) and the newly formed IG Farben in many of the most profitable sectors (organic chemistry, light chemistry, dyestuffs, and pharmaceuticals), even though the chemical industry was also highly capitalized. In this area, the DuPont directors had been irritated by the Coolidge administration's willingness to lower the customs barriers in the mid 1920s, whereas Al Smith, duly catechized by Raskob, had promised them that he would strengthen protectionism. This debate was to resurface during the New Deal. During the Hoover presidency, and as the depression worsened, Pierre du Pont and John Raskob maneuvered in the corridors of the Democratic Party to gain influence with the Al Smith and Jouett Shouse wing. Their aim was to impose a protectionist and anti-Prohibition program and also to block the political ambitions of the governor of New York, Franklin D. Roosevelt, whom they considered too favorable to labor, surrounded by radical advisors, and cryptic on the subject of Prohibition.[27]

In 1933, the cautious electoral campaign devised by Roosevelt, who made a concession to Smith and Raskob when he promised to end Prohibition and proposed a rather fuzzy economic program that was not overly alarming to big business, brought about a "marriage of convenience" between Roosevelt and a certain number of leaders of industry, among them Gerard Swope (chairman of General Electric), Cyrus McCormick (agricultural machinery), William R. Reynolds (Reynolds Tobacco), and Pierre du Pont, who gave lukewarm support to the candidate, exclusively because of his stance on Prohibition.[28]

During the campaign, Roosevelt harshly attacked Hoover for his "extravagant spending," Alan Brinkley notes, adding that this was not pure demagoguery on the part of the Democratic candidate. Roosevelt was serious about balancing the budget, and this pleased the business community.[29] And what is more, its most enlightened representatives were not hostile to a more active involvement of the federal government in economic life, provided that the business leaders preserved their preeminence. Whether or not one accepts Ferguson's typology, it is clear that big business no longer spoke with one voice by the beginning of the New Deal, unlike earlier in the century, when it presented a globally united front to the attacks of the Progressive movement.

In addition to the end of Prohibition—which was enthusiastically welcomed by Raskob and the du Ponts—the creation of the National Recovery Administration (NRA) was also favorably received by the industrial community. The NRA instituted councils for each industrial sector, to be composed of representatives of industry, the administration, and the unions, who would establish "codes of loyal competition." This idea of industrial councils had circulated in business circles since World War I—Gerard Swope had proposed it to Hoover as late as 1930—and it had been agreed to by the National Manufacturers' Association and by the du Ponts.[30] Not unexpectedly, Pierre du Pont was appointed to the Industrial Advisory Board, which supervised the NRA. But the businessmen and the Roosevelt administration were fundamentally at odds: the former wanted to control the industrial councils—keeping the government representatives in a secondary role—and set them up as authorities regulating pricing and commercial practices, whereas the New Dealers wanted to use them as instruments for economic reform. The NRA soon ran out of steam, with consumers complaining about rising prices, the unions about the stranglehold of the leaders of industry, business leaders about the administration, and vice versa. In 1935, the Supreme Court put an end to an experiment in which no one believed any longer; by that time, Pierre du Pont had already resigned from the NRA.

In the spring of 1934, a certain number of measures that favored labor and enlarged the government's authority were enacted: the Securities and Exchange Commission was put in charge of overseeing the operations of the stock market, the National Housing Act and the Wagner-Lewis Unemployment Insurance Bill were passed, and so on. A second wave of reform followed in 1935, when the Wagner Act providing for labor representation on company boards and the Social Security Act were passed with the support of a powerful union movement. At this time, the most conservative elements of the business community increasingly distanced themselves from the Roosevelt administration. A certain number of industrialists, however, among them Swope, wanted to maintain ties with the federal government and use its power to intervene for their own ends. They had understood that it was to their advantage to support a moderate reform of American capitalism, which favored social concessions within companies but also protected the interests of capitalist companies. Even before the war, major companies such as Standard Oil and General Electric envisaged their relations with the federal government in terms of a partnership, whereas the du Ponts persevered in their rigid ultraconservative stance concerning the role of the government.[31] Their alliance with the Democrats had been temporary, tactical, and linked to Prohibition and to contacts with the conservative circles around Smith and Raskob, the latter a man who saw politics strictly as a means to an end. It rapidly fell apart now, and what replaced it was an aggressive and at times paranoid anti–New Deal militancy. In this respect, the overly deterministic perspective of Ferguson, who sees a direct link between the interests of the companies studied and their attitude toward the New Deal, does not take into account the properly political dimension of the engagement of certain industrialists like the du Ponts, who were often carried away by political passion.

To be sure, by 1937, as Brinkley writes, "the idea and the reality of mass consumption gradually supplanted that of production."[32] A compromise between Roosevelt and big business became possible. And this meant that the first synthetic fiber, developed and silently woven in the laboratories during the years of political turmoil, was able to benefit from a favorable economic environment and a steady demand when it emerged. But in these early stages of the New Deal, when business leaders felt so threatened by Roosevelt and his friends, compromise was not in the cards, least of all for the du Ponts.

The Era of the Liberty League

The du Ponts' militancy assumed concrete form in the creation of the American Liberty League, officially announced on 22 August 1934, which aimed, in rather vague terms, to "oppose the radicals" and "protect the Constitution." The League's president was Jouett Shouse [also president of the AAPA], a Democrat close to Al Smith, and its governing board of six notably included Smith, Raskob, Irénée du Pont, and Stayton.[33] As Robert Burk has pointed out, the League's structure was modeled on that of the defunct AAPA, a number of whose members had agreed to serve in the new association.[34] "The Tories have come out of the woods," *Newsweek* summed up, expressing the general opinion.[35]

The Liberty League felt that the government's efforts to restart the economy were useless, indeed, worse than useless, since they slowed down the economic recovery. In their opinion, the federal government's extravagant spending and its fiscal consequences unduly burdened economic activity. Theirs was the ultra-orthodox laissez-faire discourse: both depression and recovery are natural phenomena, which no one should attempt to control—the invisible hand of the market would take care of everything. Roosevelt was ruining the fundamental nature of American capitalism by imitating communism and setting up a planned and centralized economy. Moreover, Roosevelt was a potential dictator, and the New Deal was immoral, for it turned on those whose money was the legitimate fruit of hard work and natural superiority. This kind of program was not likely to attract Americans, nor did it do much for the hapless Republican candidate Alf Landon, who had initially been rather moderate but turned to right-wing radicalism under the Liberty League's influence. The Democrats missed no opportunity to excoriate the "Du Pont Liberty League."[36]

From the summer of 1934 until the presidential election of 1936, the Liberty League progressively intensified its attacks against the New Deal, to the point where it became the noisy spearhead of the opposition, whereas the Republican Party was so damaged that it could do little more than bind up its wounds. But the League was never a grassroots association: its members were for the most part wealthy businessmen, and its political philosophy reflected overly special interests. This being the case, it was difficult to appeal to the masses to run Roosevelt out of the White House. The cartoonists, it should be added, loved it, and invariably drew the Liberty Leaguers as pot-bellied capitalists champing on cigars and sitting on sacks of dollars. As for the Republicans, they refused to deal with an association led by men who had rebuilt the Democratic Party after its defeat of 1928 and had launched extremely violent attacks against Hoover. Hoover confidentially told one of his associates of his amazement when one day he received a letter from Rascob asking him to join in his efforts. In short, the League operated in a very narrow political space.

Moreover, a few months after the operation was launched, a Senate investigation committee, chaired by Senator Gerald Nye of North Dakota, was formed to investigate war profits. Nye, one of the leading figures of the pacifist movement, had fought to keep the United States out of any engagement overseas, particularly in Europe; he felt that the relinquishment of the Monroe Doctrine in 1917 had taken place at the instigation of industrial "merchants of death" lured by the prospects of spectacular profits. In September 1934, the du Ponts therefore had to respond to a series of charges concerning their political dealings in the United States and elsewhere, as well as commercial agreements they had concluded with foreign companies such as IG Farben.[37] In the course of the hearings, Irénée du Pont, with his characteristic sense of nuance, complained that "we are no more free than the Russians." At this point, Helmuth Engelbrecht and Frank Hanighen published their book *Merchants of Death: A Study of the International Armament Industry*, which echoed the charges of the Senate committee and rapidly became a best seller.[38] *Business Week* described it as "an exhaustive presentation of the facts."[39] The hearings and the book, both of them thoroughly covered by the press, further hampered the Liberty League's first propaganda efforts.

Not that it was stingy in its efforts: banquets and meetings throughout the country, renowned publicists, and a vast headquarter in Washington—a whole floor of the National Press Club building—all testified to its vast resources. But much to Pierre du Pont's dismay, people were just not eager to join.[40] Nonetheless, between 1934 and 1936, the League stood out as Roosevelt's principal adversary, and the press gave it a great deal of attention. As George Wolfskill points out, it made the front page of the *New York Times* thirty-five times between August 1934 and November 1936, but then was mentioned only fifteen times between 1936 and 1940, the year of its demise.[41] The du Ponts had launched the Liberty League like a commercial product, with a major advertising campaign. They felt that the essence of politics lay in the presentation, in marketing, rather than in militancy on the local level.

In January 1936, the League counted 75,000 members, reaching 124,856 during the summer of 1936, and irrevocably declining thereafter, especially after Roosevelt's triumphal electoral victory.[42] According to the League's records, about $1.2 million was spent between August 1934 and November 1936, more than half a million of it in 1936 to help defeat the president in his reelection campaign. The League did its utmost for the Republican candidate, Landon, whose program closely resembled that of the Leaguers. The du Ponts contributed several tens of thousands of dollars to the cause every year—Irénée du Pont gave $ 75,000 in 1935—not to mention the sums sent to Landon: $121,530 from Irénée, $190,500 from Lammot, $111,990 from Pierre. Charles Michelson, the propaganda chief for the Democrats, referred to these sums as "dupontifical" yearly contributions.[43] In the South, the Liberty League brought out openly racist brochures, accusing the Roosevelts of inviting blacks to the White House, and also allied itself with several clearly right-wing associations, such as the Crusaders, who received financial support from Irénée du Pont, and the Sentinels of the Republic, an antisemitic organization that was also financed by several businessmen, among them Irénée du Pont and Alfred Sloan.[44]

A comparative perspective between the United States and France is interesting in this respect, for in both countries, the years between 1920 and 1930 were marked by the creation of leagues that were for the most part anchored far to the right. In the case of France, Serge Berstein distinguishes between two categories of movements designated as "leagues": on the one hand, pressure groups focused on a single cause, and on the other, political forces attacking established institutions.[45] Thus there were, on the one hand, myriads of organizations pursuing specific objectives in a specific category: examples would be the defense of taxpayers by the American Taxpayers' League, founded in 1913, and its French counterpart, the Fédération nationale des contribuables, founded in 1928 on similar premises by Marcel Large. These two catchall leagues militated for a "fairer distribution of taxes" and turned increasingly right-wing radical in the course of the 1930s, until they became very close to the Action française in one case and the Liberty League in the other.[46] But there were also leagues that promoted a more general and more explicitly anti-republican program: one of these was the Jeunesses patriotes, a movement founded in 1924 by the industrialist Pierre Taittinger and organized into paramilitary squads trained to fight back against the militants of the Left, which took to the streets on 6 February 1934. This group was comparable to the Sentinels of the Republic in the United States, an antisemitic and anti-communist movement founded in 1920, and to the Crusaders, created in 1929, which brought together activists organized into battalions that engaged in violent opposition to the New Deal. In both countries these organizations were founded and financed by prosperous businessmen (Taittinger, the perfume magnate François Coty, and the de Wendel family in France, and the du Ponts, Alfred Sloan, and James Bell, head of General Mills, in the United States), and their dream was to establish muscular, anti-parliamentarian regimes that would put social legislation to sleep, do away with the right to strike, and checkmate a presumed Bolshevik danger.

On the American shores of the Atlantic, the Liberty League was part of this nebula of far-right organizations, which between June 1935 and July 1936 came under the scrutiny of a Senate investigation committee headed by Hugo

Black, senator from Alabama and future Supreme Court justice.[47] The Black committee notably brought to light the financing of these organizations by a group of businessmen, some of whom were present in the leadership circles of the Liberty League. The two most generous donors by far were Irénée and Lammot du Pont. The du Pont brothers were at the center of a galaxy of groups characterized by their virulent opposition to the government. However, the themes they sounded came out of the register of ultra-right attitudes rather than out of any kind of American-style fascism: these characteristics made the American "springtime of the leagues" similar to its French equivalent. In other words, this was a conservative right, radicalized by the New Deal in America and in France by the leftist governments that had come to power at the same time.[48] The main difference is that in France this radicalized right found rather widespread popular support—the Parti social français, successor to the Croix-de-Feu, counted 800,000 members in 1938 and represented the principal right-wing political force—whereas in the United States, these organizations, including the Liberty League, were never able to jeopardize the hegemony of the two major parties. To this it should be added that the French leagues often had anti-capitalist overtones, unlike the American ones. From the day of its birth, the Liberty League—and this was its downfall—was too obviously associated with the interests of Big Business and a coterie of ill-tempered millionaires. Al Smith, "the happy warrior" as he was dubbed, was totally discredited by this association and left the public stage by the back door.

The fact is that the du Ponts and their friends never understood that for the vast majority of Americans, the enemies were not Roosevelt and the income tax but the depression and poverty. In the correspondence of Pierre, Irénée, and Lammot there is not one word about the suffering of large numbers of their compatriots, not one note of compassion, not one remark about the country's social conditions, and nothing but lamentations and unending recriminations against the burden of taxation and the politicians.

The du Ponts might have aligned themselves with the moderately anti-Roosevelt positions of the National Association of Manufacturers (NAM), which dovetailed quite well with their interests.[49] Yet, moved by what one must call, for lack of a better expression, a political passion, Irénée and Lammot ventured much beyond that position, very much to the right, to the point of flirting with the racist and antisemitic far right. In so doing, they further compromised the image of their firm, which had already been severely tested by the Senate investigation committee on war profits. "The Americans' favorite family," as Charles Michelson has put it succinctly, had become "public enemy number one."[50]

And yet Americans could have clearly distinguished between the du Pont family and the DuPont firm. After all, to buy a can of DuPont paint was not to agree with the political stance of the family. But it seems that the marginal political positions of the directors did harm the company.

According to a Gallup poll, only 20 percent of Americans had a good opinion of DuPont, as opposed to 80 percent for General Motors or General Electric.[51] Although one should not take this poll too literally, it does testify to a certain concern about possible negative effects of politics on business. This concern was not unfounded: Roosevelt would occasionally make ironic remarks about the standard bearers of a bygone era, the time of a high capitalism that was deaf and blind to social demands, and certain members of Congress demanded that no Army or Navy contracts be given to DuPont.

The response of the chemical firm turned out to be well thought out and well implemented. Instead of engaging in a suicidal test of strength with the government, DuPont's directors decided to promote their firm by means of the products it made, and to make sure they reached a mass market. To carry out this policy, they hired a renowned public relations specialist, Bruce Barton, known for his book depicting Jesus as a brilliant businessman who had started a formidable business with twelve partners.[52] Barton had little trouble making the DuPont directors understand that they had to place the emphasis on products for the civilian mass market in order to erase the negative image of the "gunpowder trust." A new slogan concocted by his agency (Batten, Barton, Durstine & Osborn) came out in October 1936: "Better Things for Better Living . . . . Through Chemistry." From then on, this slogan was systematically inserted next to the firm's logo—as it is to this day, except that "Through Chemistry" was discreetly removed in 1972. In addition, a radio program, *The Cavalcade of America,* was broadcast every week on CBS in order to highlight the benefits of chemistry and the patriotism of the company. This was not simply a public relations campaign orchestrated by a publicist, it was the fruit of a thoroughgoing reflection on the structural changes of the American market. In accordance with the publicist's recommendations, the firm's booths at the major fairs featured plastics, artificial rubber, paint, lacquers, and varnish behind gleaming panes of glass, whereas gunpowder was kept in the background and symbolized only by a box of shells for hunters placed next to a game pouch. In this connection, Meikle cites a poll conducted by the students of Southern Methodist University about the DuPont booth at the Texas Centennial Fair in the summer of 1935.[53] According to this poll, the visitors interviewed declared themselves favorably impressed by DuPont's civilian activities and did not think that the company "needed another war," which was, of course, exactly what DuPont wanted to hear, so that this poll must be taken with a pinch of salt. Nonetheless, it does show how much importance Barton and his clients attached to a communications strategy that would be in tune with the potential customers. Barton also advised the directors to keep a low profile as far as their tense relations with the government were concerned. After all, he told them in effect, there is no point in rowing against the current of a rather pacifist public opinion that has embraced Roosevelt. Forget about your cutting statements; what you need to do is to stress the achievements of your chemists and the benefits of chemistry. The launching of nylon in 1939 provided an excellent opportunity to do just that.

Nylon and Political Compromise

DuPont's directors used the occasion of the commercial debut of nylon to proclaim their consumerist credo. In a highly characteristic speech, Henry B. du Pont, the vice president of the firm, asserted before an audience of Michigan industrialists on 24 January 1940 that "through technology and mass production industry has conclusively shown, over a period of decades, that it has produced jobs for millions of people and raised the standard of living of all classes. Technical developments of the past few years have laid the foundation for additional millions of new jobs."[54] Walter S. Carpenter, who became chairman of DuPont in 1940, echoed his words when he asserted that "the best way to increase the size of the cake and to a fair distribution of income is to invest into industrial research, cut down the costs and the prices of the products of industry by research."[55] This was said when American liberalism had just accomplished its great political molting by accepting high capitalism and the existing structures of the economy and by placing mass consumption at the center of economic thought.[56]

This last perspective was new in the mouths of politicians, even though over the preceding twenty years, economic growth had fundamentally been driven by demand for consumer goods.[57] The first years of the New Deal had been marked by government efforts to reform production. But now, precisely at the end of the 1930s, the idea of restarting the economy by means of consumption had made its way into the highest councils of government, and in this area, a tacit agreement with the leaders of big industry was possible. To be sure, the other mainstay of Keynesian policy, namely, the idea of public spending to counteract the cyclical downswings of the economy, remained a point of disagreement until after World War II, but the anti-capitalist rhetoric of the New Dealers had definitively been put aside.[58]

In this more conciliatory environment, nylon assumed its full significance. Raymond Clapper, a renowned editorialist, exclaimed in the *New York World Telegram* of 17 January 1939:

> These du Ponts! For several years they spent thousands of dollars on anti-New Deal propaganda, subsidized the Liberty League, appeared by the dozen to applaud Al Smith, and the bejewelled royalist anti-Roosevelt meeting here a couple of years ago financed court obstruction to social and economic reforms—all on the ground that Roosevelt was ruining private enterprise and making life impossible for businessmen.
>
> And now it is the du Ponts who, by their genius for initiative and industrial adventure, are demonstrating to businessmen that aggressive brains can still produce business, even while that man Roosevelt is in the White House.
>
> Today, after six years of Roosevelt, Du Pont executives, as I gathered at luncheon with some of them, still think he is ruining the country, but they are employing 52,000 men and women now as against 42,000 in 1929 at the peak of Hoover's prosperity. In the face of all the New Deal measures which they so expensively said would destroy private initiative, the du Ponts themselves are providing the most thriving answer to their own gloomy political views.[59]

Nylon once again brought DuPont to the front pages of the newspapers, albeit in a more favorable manner than usual. In the spring of 1940, with cannon thundering in Europe, and the attacks of the Nye Committee not yet forgotten, the very promising success of nylon further strengthened the Executive Committee's conviction that the civilian market was not subject to "political reverberations."

Moreover, the du Pont brothers stepped out of the limelight. Lammot du Pont resigned as chairman of the company, and Pierre and Irénée followed as chairman and vice chairman of the board on 20 May 1940. This allowed them to devote themselves full time to the presidential campaign of Wendell Willkie, the Republican candidate and director of an electric company in the South, who had become known for crossing swords with the Tennessee Valley Authority and its administrator David Lilienthal.[60] Lammot was replaced as DuPont's chairman by Walter Carpenter, the first chairman of the firm who was not a du Pont, and politically rather more discreet than Lammot. And as a last symbol, Ethel du Pont, daughter of Irénée, married Franklin Roosevelt Jr., the son of FDR and Eleanor, in 1937. The radical writer Ferdinand Lundberg considered this union a sign that Roosevelt had rallied to high capitalism.[61]

On the brink of another war, the retiring du Pont brothers left behind a considerable legacy in the form of a company that had achieved its strategic conversion by betting successfully on the expansion of consumer spending. Nylon was the culmination of their technological and organizational efforts. At the same time, they had held on to a nineteenth-century vision of politics and public action, which had showed itself to unfortunate effect in the Liberty League, although it was to be rapidly erased when World War II broke out and when new actors took center stage. For the moment, these actors were still backstage.

Company Culture

Directors and Employees

How did the company and its employees fare? After all, the political considerations outlined above concerned the du Ponts and not their company as such. There can be no question, of course, of looking for a clear connection between the political opinions of the family members who served as directors and those of the company's engineers and managers, nor even between political choices and the conduct of business.

Historians of the Progressive tradition, including the revisionists of the New Left, have assumed that there is an evident relation between politics and economics, and that the socioeconomic interests of the dominating classes are reflected in their political machinations. In his day, the Progressive historian Charles Beard had already endeavored to demonstrate that the Fathers of the Constitution had been motivated by their special interests.[62] Later such revisionist historians as William A. Williams and Gabriel Kolko adopted a similar stance, claiming that the foreign policy of the United States and the domestic policies of different governments were dictated by the interests of big companies.[63] Here politics is reduced to a superstructural modality of economics.

In this light, the DuPont directors were seen simply as employers seeking to reduce production costs as much as possible, who had no use for social legislation, and whose social Darwinian philosophy and paternalist stance barely concealed their greed. In attacks reminiscent of those the leftist parties launched against the "two hundred families" in France in the 1930s, the communist *Daily Worker* and left-wing Democrats gleefully branded the du Ponts one of the sixty "most powerful and most maleficent families" in the United States.[64] The du Ponts presented an almost perfect target for radical critics: not only were their political opinions quite ultraconservative and anti-union, but they lived lavishly, in luxurious mansions—among them Granogue, Nemours, Saint-Amour, and Longwood in the Wilmington-Philadelphia area, Alfred du Pont's Montpelier (the former home of President James Madison) in Virginia, and Irénée's fabulous Cuban residence, Xanadou. They were frequently called the American Krupps by the press, and sometimes "the explosion kings of the capitalist world."[65] When Gerard Colby, assistant to a member of Congress from Delaware, published a highly polemical book in 1974 titled *Du Pont: Behind the Nylon Curtain*, attacking the du Ponts violently and seeking to show that even in the early nineteenth century, they had represented all the worst aspects of American capitalism, DuPont evidently had large numbers of copies bought up and destroyed.[66]

Colby never demonstrates the link between the political and the economic levels; it is assumed and implicit. Yet Alfred Chandler's studies convincingly show that the company's managers and engineers had considerable room for maneuver in their day-to-day decision-making and were not simple pawns who passively followed the instructions handed down by their boards. The running of a business must therefore be dissociated from the behavior of the directors (with which, moreover, Chandler does not concern himself).[67] According to Chandler, stock market speculation and the political maneuvers of prominent directors had little impact on the daily life of companies like DuPont, which depended above all on the work of a multitude of anonymous employees. Furthermore, as Olivier Zunz has shown in a social-history perspective, the mid-level managers who helped bring about the rise of these big companies did not necessarily have the same interests and the same philosophy as their directors, even if they shared quite a few of their values.[68]

But, conversely, one must not postulate an absolute, watertight separation between these two areas either, as business historians are wont to do.[69] For Chandler and his followers, nothing interferes with the course of business and with the relation between the company and the market. The manager, a *homo oeconomicus,* is assumed to act at all times in keeping with rigorous economic imperatives, undisturbed by extraneous concerns. The management and the strategy of the company are ruled by considerations that have nothing to do with the political opinions of directors or shareholders: hence the Chandlerian historian Charles Cheape has entitled his biography of Walter Carpenter (who became chairman of DuPont in 1940) *Strictly Business.* This interpretation too seems limited, for while there is no denying that the goal of every business is profit and the pursuit of the most advantageous way possible to sell its products, the economic and social importance of a large organization cannot be reduced to this commercial dimension alone. A great corporation is a sociocultural entity made up of tens of thousands of employees, who work collectively, irrigated by instruments of cultural diffusion, such as company newsletters, meetings, and the training sessions of which modern companies are so fond. It is also a political entity, not only in a narrow sense of power relations within the organization, but in the sense that it brings its weight to bear on all social relations and on political choices at the local and national levels.

In this connection, the familiar and rather overused expression "company culture" may deserve to be reconsidered. This notion has become common in the discourse of management, notably thanks to the classic study *In Search of Excellence* by Thomas Peters and Robert Waterman, which advances the idea that in prosperous firms, there is a more or less definable set of factors that motivates the actors and affects their actions.[70] In recent years, many articles and books have debated the coherence between values and behavior, the relation between the culture and the efficiency of organizations, and the existence of several cultures in one and the same organization. The advantage of the notion of company culture is that in the analysis of an organization, it brings out other factors than just structure. It is therefore used with reckless abandon by management specialists, who take it to signify that companies usually have their own culture that strengthens cohesion and a sense of belonging among the employees, particularly at the upper echelons—unlike in certain large and impersonal organizations, where social relations supposedly are purely instrumental. The misunderstanding no doubt arises from the fact that this notion is not so much descriptive as prescriptive: management theories are not meant to report on reality but aim to be useful to managers. Hence their explanatory value is usually small—and their life span limited. Should we therefore get rid of them? Social scientists could reappropriate this notion, provided they do not consider it proven and work out its precise meaning: if there is such a thing as company culture, it must be demonstrated and circumscribed.

The HOBSO program

DuPont had early on put into place training programs for both workers and supervisors that featured practical advice alongside a more political discourse. Every new employee had to participate in a program called "How Our Business System Operates" (HOBSO for short), which was intended to reinforce cohesion and inculcate certain of the firm's values in everyone. As far as the workers were concerned, it was important to harden them against any temptation to favor independent unions and to urge them to join the company union, the Works Council.[71] The brochures distributed to newcomers asserted the principles of laissez-faire according to the most orthodox ideology, warned against the "hegemonic" intentions of the government, and aimed to turn the newcomers into "Du Pont men."

The training for new engineers was even more crucial, of course, since their whole-hearted support of the company's objectives was necessary to its proper functioning. The engagement and the motivation of engineers and managers was a long-standing issue.[72] Since this engagement could not be measured by strict Taylorist criteria, such as punctuality and productivity, these employees themselves had to be convinced that their work was of interest to themselves and to society. Material advantages were not enough, and the HOBSO program was designed to give an ideological content—or to confer a "spirit," to use the formulation of Luc Boltanski and Eve Chiapello—to the exercise of the profession of chemical engineer at DuPont. But then the newly minted graduate engineers from the universities probably already had some ideas about this subject.

In the late 1930s, newly hired chemical engineers were expected to take a course entitled "The Engineer in a Free Economy,"[73] given by Charles Stine himself, the vice president of DuPont, who became president of the AIChE after the war. Stine began by explaining to them that the United States owed a great deal to its engineers, and that thanks to the chemical engineers, the country's chemical industry had now become first in the world. So what is a chemical engineer? First of all, he is a "practical man":

Unlike his cousins in the categories of the chemist and the pure physicist, our engineer must not dream incorrect theories. The engineer dares not indulge in the amusing intellectual pastime of writing long and faulty dissertations about a world of functional natural laws. It is to this functioning and integrated world of natural laws that the engineer's devices and structures must strictly conform. In a word, they must work.

This does not mean, of course, he kindly adds, that the chemical engineer is stupid. Developing nylon, the speaker pointed out, had involved dealing with multiple problems. Could they have been solved by men who were "slightly illiterate" or "comfortably ignorant"? If the chemical engineer was a "practical man," it was in part because he lived in a "free economy" and "could not survive in an economy controlled by the government."

> The whole setup operates to crush him, to negate his efforts, to rob him of his independence, to cramp and channelize his thinking, to subject him to political controls and pressures. Would it be the same in a state-controlled United States as it has been in every state-controlled nation? It would, if history means anything, it would have to be, inevitably.
>
> It's easy to see what such control would do to our country. It would mean stagnation, and rot. But it's also easy to see what it would do to the practical man. It would be the end of him. The engineer cannot exist save in the unregimented freedom of the republican form of state, and the state moves forward only by encouraging the practical man to do his utmost.... Oddly enough, many people would like to see such a setup in the United States, and for its ultimate realization they are working fanatically.
>
> But I hope that no engineer will ever be misled by this type of propaganda for he, far from being nonplussed by the irrelevance of facts, is very much disturbed by it, and his intellect enables him to spot those irrelevancies in the vaporings of the government-control advocates.

Stine continued with a warning against the "true reactionaries," namely, the socialists and communists who want to "take over the factories." And so he concludes:

> The real progressives are those who refuse to believe that we have reached our limits, who realize that to stand still is to go backward, who know that the greatest security for all lies in moving forward.
>
> Truly practical men know these things. They've spent their lives practicing

that creed, so that their experience proves what their intelligence tells them. That is one of the most hopeful things I know about the future of this country, for the practical men in all fields have a way of prevailing. There is much to do. There are frontiers as yet undiscovered. The practical man will be in the forefront of the drive to open these new frontiers.

The engineer, who is the practical man, repays the economic opportunity and the political freedom given him by our system by enlarging and multiplying the economic opportunities. This increase in economic opportunity is what gives solidarity to our free economy.[74]

New employees also attended lectures on the company's history, on the need to speak proper English and to go regularly to the library in order to keep up with technical reviews (Crawford Greenewalt remembered reading German chemistry journals),[75] and on communicating on a regular basis with salesmen, workers, and finance managers. The training program thus mixed political considerations—which, while tinged with anti–New Deal attitudes, were nonetheless fairly cautious—with commonsense advice on how to succeed in the company. Taken as a whole, it outlined the main elements of the company culture that the directors wished to spread among the managers and engineers. It was characterized by two main aspects.

The first of these was that DuPont's political values were more conservative and more militant than those of most other major companies of that time. These values were directly linked to the political philosophy of the du Pont family. The fact that Lammot du Pont's deeply trusted right-hand man Charles Stine personally delivered the political lecture shows that it was considered very important. To be sure, Stine did not expressly mention the political skirmishes that pitted the du Ponts against the New Dealers, but there is no doubt that everyone was aware of them. Like all other Americans, the engineers read the newspapers and knew about the attitudes of the eponymous family. It should be added that certain leaders of the Liberty League wanted the association to have access to the roster of DuPont employees, so that brochures and membership applications could be sent to them. Lammot du Pont was cautious enough to refuse, as did Alfred Sloan for General Motors. Yet the DuPont directors expected their upper-level employees to show a certain allegiance to a set of conservative political values, perceiving this as a sign of personal fidelity and loyalty to the company—even if they did not expect those employees to join the League.

When the occasion presented itself, DuPont's directors did not hesitate to distribute brochures with anti-government overtones. Thus, in 1935, at the time of the Social Security vote, an information sheet about Social Security signed by F. C. Evans, the director of the personnel department, was distributed to all employees. It combined factual information about the new legislation—regarding both its insurance and its welfare provisions—with considerably more political statements to the effect that the American taxpayers would have to pay between three and four billion dollars a year for retirement pensions and unemployment insurance. "Experience shows that neither prosperity nor economic security are guaranteed by this program."[76] The company directors seem to have had no illusions, however, about their employees' sympathy for the conservative discourse they espoused in public. They asked the heads of the various departments to report on the reactions to Evans's writings. The report records contrasting sentiments, ranging from "favorable" to "indifferent" and above all to "mixed comments" and "arguments received badly," "unfavorable reactions to political subjects," "only one point of view presented," "political subjects should be avoided."[77] These somewhat fuzzy remarks are nonetheless significant, considering that it was not in the interest of the department heads to make too much of these defiant reactions. It therefore appears that the employees' reaction was on the whole negative, and that the political speeches of the company directors failed to hit the mark, particularly among the workers. This gave rise to certain convoluted explanations noting that it was very difficult to write for "ordinary workers" or that the articles had not been distributed beyond the foremen. It is probable that the reaction of the engineers was more favorable (although the report does not specify this), for their social position and their career interests would motivate them to respond according to their employer's expectations. The following year, two articles in the same political vein by DuPont's vice president Jasper Crane were rejected by the ad hoc committee of the Personnel Department, which called them interesting but recommended that they not be distributed.[78] It is likely that the unfavorable opinions recorded the previous year had made the department heads more cautious.

Secondly, the program explained that newly recruited chemical engineers should make their mark by their work. You are a university graduate, they were told, but that is not enough to make you a chemical engineer. "The schools do not train chemical engineers; they prepare you for the profession of chemical engineer."[79] For the moment, you don't know very much, you will have to work hard, and your rise in the firm will depend on how well you do. The initial devaluation of academic competence was designed to stress the importance of on-the-job training and of the internal rules for promotion. "The diploma is not everything," and the ultimate verdict was delivered by the company, rather than by the university, whose modes of functioning were not entirely controlled by the industry, for all the influence it exerted in these matters. Revealingly, Crawford Greenewalt reiterated that in college, one is not judged by the success or failure of a project, adding: "I must say that some of this self-discipline would do the universities and the research people a lot of good."[80] By this he meant to convey that the career of a DuPont engineer was subject to other criteria than those of the university, where the rules of the market were not involved to the same extent, and where social relations were also different.

A Strong Group Identity

But a few lectures are not enough to forge a strong group identity: such a bond is built up gradually, and social identity involves a dialectic between what is expected of those concerned and their response to those expectations. It is not easy to gauge to what extent DuPont's managers and engineers were open to the standard arguments of the company directors. However, their interviews and extracurricular activities, which were almost exclusively connected with DuPont (membership in the Henry du Pont athletic club of Wilmington, and the Country Club, for instance, or various cultural activities offered by the company) indicate a very strong sense of belonging to the chemical firm, and no doubt approval of its conservative values. It stands to reason that the normative messages from the board met with the general approval of the engineers, considering that such messages fulfilled their expectations and responded to what Boltanski and Chiapello call the "need for justification."[81]

Beyond their careers and the material comforts provided by their salaries, the engineers felt the need to legitimize their profession in terms of the common good. They did not want to lose sight of the underlying motivations of their professional commitment, which were centered on the benefits provided by technological progress. Chemical engineering was valued for allowing all Americans to have access to sophisticated consumer goods such as nylon. In a context of economic depression, marked in politics by the rise in the government's capacity for intervention and in cultural terms by a more or less explicit questioning of the market economy, DuPont engineers saw themselves as diligent agents of capitalism. What Boltanski and Chiapello call "the spirit of capitalism"—that is to say, the ideology by which one justifies a commitment to capitalism, "in which the salary constitutes at most a reason to stay in a job, but not to devote one's life to it"[82]—was centered on general values of happy consumerism, acceptance of which peaked in the 1950s, before they were once again called into question at the end of the following decade. By then the chemical engineers had come to be seen not so much as exemplary figures of modernity than as its sorcerer's apprentices. But at DuPont in the 1930s, the discourse of capitalism was reinforced by a set of themes celebrating the market and economic liberties, which were more political than at other companies. The setup seemed to suit the company's engineers. In any case, the archives and especially the interviews do not furnish any indications of dissatisfaction. Working for DuPont evidently seemed to be sufficiently attractive in terms of ideological justification, material advantages, and career prospects.

Yet one should not think of the DuPont engineers as elites endowed with economic, social, and cultural capital that would make them the equals of the firm's directors. While it is not possible to identify the social origins of all these engineers precisely, the available documents do indicate that they were sons of the upper to lower middle classes, and that becoming an engineer represented a definite rise in social status for them. As they themselves noted, many had grown up in small towns in Pennsylvania, Ohio, Michigan, Illinois, Wisconsin, and Iowa in families of northern European origin with well-established local roots, often thanks to business ventures, and whose incomes were sufficient to send their male children to college. Some mention summer jobs during their college years, and all speak of the Great Depression as an important time in their lives, during which they experienced fear of unemployment and dark days.[83] Their preoccupations matched those of most other Americans, rather than those of an elite shielded from the crisis. Until World War II, their social position was not firmly established. Many also talk about the origin of their interest in chemical engineering by emphasizing their passion for tinkering and physics and chemistry experiments, although they do not mention any specific intellectual motivation. The picture that emerges is that of relatively undistinguished average Americans in a mid-level profession.

A priori, one should neither deny nor presuppose the relation between economics and politics. Rather, that relation should be considered in cultural terms, without mechanically indexing political opinions to socioeconomic attributes; the organization should be seen as the site for the diffusion of a given culture. Beyond the initial training of the newcomers, the social and political culture of the DuPont employees was characterized by two outstanding features: a highly developed pride in the company; and the influence exerted by the directors, specifically the du Pont family.

The pride in the company started with its original work. Making gunpowder was dangerous, accidents were frequent, and working for DuPont was the sign of a certain courage: it took a DuPont *man*. As one Italian workman put it in his hesitant English, "I worka da boom boom factory. Son-a-bitch job!"[84] One geographic factor reinforced DuPont's special identity, for its industrial facilities were located at considerable distance from urban centers for evident reasons of safety. The original site of the firm is quite isolated on the banks of the Brandywine River some ten kilometers from the center of Wilmington. The employees lived close to their work, among themselves and without connections to the region's socioprofessional network, and this accentuated the special identity of the firm. In the nineteenth century, for instance, to protect trade secrets, the company continued to use French as its "technical" language for several decades after its establishment by Eleuthère Irénée du Pont.

Accidents and work-related illnesses resulting from the handling of toxic substances, such as the fatal poisoning of several dozen workers by lead tetraethyl in the 1920s and cases of liver cancer caused by certain tinctures in the 1930s, demonstrated that the chemical industry was more dangerous than many others. The brochures about the company that DuPont distributed to engineering students at colleges and universities made it clear that working in a chemical plant demanded "a strong constitution" and "plenty of staying power."[85]

Correlated with this image of heavy industry and tough work, sexual discrimination was more prevalent in chemical engineering than in other specialties. Of the 7,500 female DuPont employees in 1940 (or 15% of the total of 50,000), only 5.5 percent (or 412 women) were college graduates, and only 3 percent (225 women) had advanced degrees. Of these 400 female college graduates, roughly 200 were laboratory technicians (most of them in the Rayon Department) and about 150 worked in the medical division, while some 50 women held positions as librarians in the company's twenty-seven libraries.[86] As far as we know, no female chemical engineer was hired by DuPont before the late 1950s. According to J. O. Maloney's statistics, only two women obtained degrees in chemical engineering between 1902 and 1959.[87] This pronounced gender inequality may have contributed to the close relationship between chemical engineering and its military clientele in the 1940s and 1950s.

A more diverse staff would have provided a breath of fresh air, opened the company up to new social and cultural influences, and brought in leaders of different backgrounds, but the Executive Committee insisted on giving the directorships of new plants and new branch offices to DuPont men who could be counted on to pass on the same message to their employees. Brochures celebrated the firm's accomplishments. DuPont *Magazine*, launched in 1913, highlighted the model employee, who lived by the classic values of hard work and exemplary loyalty to the company. The firm's collective memory was kept alive, and the employees knew that their company had been founded for the purpose of providing gunpowder for the Republic, that the North had won the Civil War with DuPont powder in its cannon, and that production feats had been accomplished during World War I.

As in other companies, engineers had become a key element in DuPont's development by the beginning of the century. Interviews with these engineers conducted by Chandler and his assistants in the early 1960s show how deeply they identified with their company. The hiring of chemical engineers from 1920s on did not change this. These men were loyal to Du Pont, where they spent their entire careers (except for a few, like Allan Colburn, who went back to academia; but this was always done with the consent of the firm, with which they remained in touch as consultants and teachers of future DuPont engineers). The company was so large and so diversified that an ambitious engineer could spend his entire career there: first as production or research engineer, then as office manager, initially in a branch and eventually at company headquarters in Wilmington.

Moreover, their attachment to the values of law and order, their conservative political philosophy—which had grown out of the family traditions of old-line Protestantism in the middle classes of Pennsylvania and adjacent states—and their esprit de corps designated them as a very special group: in this respect they were not unlike army officers. Hence, perhaps, as we shall see, their collusion with the military, reinforced by the fact that some of them were reserve officers trained in the ROTC units that had first been set up at MIT in 1911 and then nationwide over the next years.[88] When Greenewalt came to MIT in 1918, he joined the Student Army Training Corps (SATC), which had just been created.[89]

It would also seem that certain of the company's departments, such as the Department of Chemical Research, favored hiring WASP engineers, rather than Jews, for example. Hounshell and Smith report that the German Jewish chemist Hermann Mark (the expert in polymers) was not hired in 1933 because of the antisemitism "of certain leading members of the department" (most likely Bolton).[90] Needless to say, African American engineers were not found there either.

In exchange for this fidelity to the firm, DuPont paid its engineers higher salaries than other companies and supplemented them with substantial bonuses and profit-sharing arrangements. In 1912, 1,440 of DuPont's employees, representing more than half the total number of shareholders, owned shares in the company. The thirty managers at the top of the hierarchy received the lion's share.[91] Moreover, a series of bonuses was designed to stimulate and retain the best engineers and managers. "A" bonuses were distributed every year to employees who had contributed to the improvement of a process or a product (Greenewalt received one of these in 1923); and "B" bonuses rewarded managers according to the company's global performance.[92] Instituted by Pierre du Pont when he came to the company in 1903, the bonus policy was systematized by his successors, in particular by Walter Carpenter during his tenure as chief financial officer. Before World War II, the compensation policies of DuPont and the similarly run General Motors were among the most sophisticated in the American business community. However, many employees, including certain managers, were critical of the bonuses and complained that that they were governed by a policy of favoritism. The general manager of the Ammonia Department said frankly that "practically all of our men think that the 'A' bonus does more harm than good."[93]

Medals were also distributed to faithful employees: five years in the service of DuPont earned one a gold-plated service pin, which would be embellished with a gold star five years later, and with two, then three stars ten and fifteen years thereafter. A twenty-five year seniority was rewarded with a medal in white gold decorated with four diamonds.[94]

Since the company offered such perks and symbolic markers, and engineers were in any case recruited primarily from universities with which it had agreements, DuPont acquired assiduous, loyal men who were deeply devoted to their work. As Roger Williams summed it up: "Du Pont is my life."[95]

Another aspect of DuPont's culture was that the company was still directed by the descendants of the founder. The major transition from family capitalism to financial capitalism, which took place in the United States earlier than in Europe, had been negotiated in an unusual manner at DuPont, since the eponymous family had kept control over it, thanks essentially to Pierre du Pont's strong personality and remarkable management skills. One of the essential features of modern capitalism, the separation of the control of capital from the management of the enterprise, did not apply at DuPont. The cousins and then the brothers du Pont were both the directors and the principal shareholders of the firm, something extremely rare in the history of American high capitalism, aside perhaps from the Ford Motor Company.

The consequences of this familial dimension were quite concrete. In the nineteenth century, the company's directors sometimes did not hesitate to get their hands dirty and work alongside their employees. In 1884, Lammot du Pont Sr. was killed by an explosion when he and his chief engineer rushed to the aid of workers who had mishandled material in the nitrate storage facility. The du Ponts lived very close to the workshops, and their company was the archetype of the family enterprise—indeed, before 1902, the boundaries between the company's operating funds and the directors' pocketbooks were not altogether clear. Pierre and his two brothers, Lammot Jr. and Irénée, ran the company until 1940, successively and each in his turn occupying the chairmanship of the finance board, the general chairmanship, and the chairmanship of the board of directors. The three brothers also controlled the holding companies, Christiana Securities and Delaware Realties, that held 30 percent of the firm's stocks. Until World War II, Pierre du Pont wielded considerable influence over the company's policies in general, imposing some of his choices even when the Executive Committee disagreed. In the early 1930s, for example, the committee yielded to Pierre's desire to transfer DuPont's New York offices to the newly built Empire State Building, which he had commissioned along with his old crony John Raskob, and which was largely vacant at the time owing to the depression.[96]

The proximity of the du Ponts, in conjunction with their political commitments, contributed to the spread of a conservative message tinged with a certain persecution complex throughout the organization. Its directors felt that they had been the object of repeated and unjustified attacks by the government and by Congress, and believed that the politicians, being irresponsible demagogues by nature, held a particular grudge against DuPont. It is not easy to judge whether this mind-set had spread to the engineers and managers, but one can assume that the company was powerful enough to convey a minimum of cultural baggage, a "least common denominator," to its mid- and upper-level management. It included first of all the pride of being part of DuPont and of having been chosen by it; the engineers interviewed "never knew" why they had been hired, thereby implying a kind of lasting astonishment that they had been selected and then promoted by such a firm; correlated to this was an identification with the most flattering representations of the profession: DuPont was thought to employ the best chemical engineers, which it first had to identify in the forcing grounds set up by professor consultants at the universities. Another aspect of this culture was loyalty to the company, to which these men subordinated their social relations (the regional meetings of the Association of Chemical Engineers refer to gatherings open only to DuPont engineers and to parties given in honor of a specific DuPont man), and to which they quite possibly adjusted their conservative values. The company expected its engineers to share in its political culture (political conservatism, Protestant ethic) to some extent, although it is difficult to deduce precise political positions from this. The concept of company culture, then, refers here to a set of values suggested by the strongly differentiated identity of the company under study. It should be added that this did not make the DuPont engineers radically different from their colleagues elsewhere. Arthur Morgan, a hydraulics engineer who founded Antioch College anew in the 1920s before becoming the first director of the Tennessee Valley Authority, sadly noted that the engineer tends to reflect, in an uncritical way, the social attitude of his employer.[97]

At that time, in the prewar United States, the cultural and personal characteristics of engineers and managers had not yet become an issue. In the 1950s, by contrast, a new fear came to light, that of having men entirely shaped by their company, becoming interchangeable look-alikes in their dark suits, white shirts, and dark ties. In 1929, in a book written for young engineers, Daniel Mead advised avoiding all sartorial eccentricities, "for showy clothing denotes inappropriate idiosyncratic behavior that must be eliminated."[98] The sociologist C. Wright Mills referred to this attitude in his book *White Collar*, in which he deplored the uniformity of these masses of employees, whose lives were reduced to performing work without intellectual horizons for their companies.[99] *Du Pont Magazine* published articles encouraging employees to express their individual tastes and temperaments, and Greenewalt himself, by that time president of the company, published one in which he declared that he himself did not wear the stereotypical gray flannel suit, and that his staff were under no obligation to adhere to strict social codes, saying:

> I realize that it is popular to regard the large business unit as a machine in which human tolerances are held within precise limits and some All-Seeing Eye charts the manners, dress, and political views of each candidate for advancement. One of the magazines devoted to business has the curious conviction that the wives

of business executives are screened critically as part of the criteria for their husbands' promotion.

All this of course is pure fiction, at least according to our experience. The Gray Flannel Suit is a pretty superficial symbol after all (I don't own one), and among my most valued associates taste in dress covers a pretty broad pattern. The same goes for personal habits and enthusiasms. As for wives, I can report that among my close co-workers there are a number whose ladies I have never seen.[100]

Photographs taken at meetings of DuPont engineers and upper-level managers of that era do not reveal a great deal of sartorial eccentricity and seriously qualify Greenewalt's optimistic affirmation, but the point is that his words express an interest in some relaxation of the strict social codes that no one thought of questioning before World War II. Greenewalt also encouraged his employees to take an interest in the outside world: become cultivated, he told them in essence, read, listen to music, so that you will no longer be stereotyped as boring people stamped out of the company's mold. Greenewalt's personality and his "aristocratic" background contributed to distinguishing him from the petit-bourgeois conformism of his subordinates.

However, it seems to me that one fact did have a "centrifugal" effect, and that is generational change. This was discreetly underlined by Greenewalt when he stated as an obvious fact that "we were much younger than they [the company's directors]."[101] This cautious qualification was perhaps Greenewalt's way of distancing himself from a certain political archaism of the directors, and of attributing it to their age—a way of saying that they, the Young Turks, would have done things differently. And the fact is that the war would give them the opportunity to demonstrate a new way of handling politics, to develop a new type of relation with the public sphere, and to engage their company in what was not yet called the military-industrial complex, thereby bringing it politically up to date.

Making black powder in a DuPont workshop around 1895. The structure was wood, the implements were copper, and there was an emergency exit (far right) in case something went wrong. DuPont powdermen took their time; their trade was dangerous, but they knew it well. No need for chemical engineers here. Courtesy Hagley Museum and Library.

Julian Hill, a chemist who worked with Wallace Carothers, shows in a 1946 reenactment how he extracted the first nylon filament from a test tube eleven years earlier. DuPont was already aware of the legendary dimension of nylon and spread the gospel of chemistry. Courtesy Hagley Museum and Library.

Artist's rendering of the DuPont pavilion at the 1939–40 New York World's Fair, designed by Walter Teagle. The tower, filled with liquids, was ninety feet high. The message was that a brave new world was in store, thanks to the achievements of chemistry. Postcard from the author's collection.

The DuPont exhibit at the 1939–40 New York World's Fair featured elegant young women dressed in nylon clothing and stockings "more delicate than silk." Here a photographer captured them posing in front of a banner reading "Nylon: A new word and a new material for all mankind" and surrounding a chemist and his knitting machine. He, of course, occupies the dominant position; the "nylon girls" flash him admiring smiles. Courtesy of the Hagley Museum and Library.

Movie star Betty Grable sets a patriotic example by auctioning her nylon stockings for war purposes. They fetched $40,000. Courtesy Hagley Museum and Library.

Reactor D under construction at Hanford in June 1944. The reactor is in the middle of the picture, with the long building of the chemical separation process in the background. Courtesy Hagley Museum and Library.

In 1946, the key men of the Manhattan Project met to celebrate the appointment of physicist Arthur Compton to the presidency of Washington University in Saint Louis. From left to right, seated: General Leslie Groves, Vannevar Bush, Enrico Fermi, Colonel Kenneth Nichols, George Pegram, Lyman Briggs. Behind: Charles Thomas, James Conant, Arthur Compton, Eger Murphee and DuPont's Crawford Greenewalt. Many of these men would be deeply involved in the academic-military industrial complex. Courtesy Hagley Museum and Library.

TIME
THE WEEKLY NEWSMAGAZINE

DU PONT'S GREENEWALT
Cellophane, nylon, a wrinkleproof suit—and the H-bomb.

DuPont's president Crawford Greenewalt appeared on the cover of the 16 April 1951 issue of *Time* in the middle of the Korean War, surrounded by objects symbolizing the company's activities. © P.P.C.M.

CHAPTER FOUR

The Forgotten Engineers of the Bomb

In early August 1945, the blinding flashes that devastated Hiroshima and Nagasaki instantly propelled the Manhattan Project from the deepest secrecy to the top of the news. One week after the atomic explosions, the portrait of the charismatic J. Robert Oppenheimer was on the cover of *Life* magazine. The traditional account of the history of the Manhattan Project gives star billing to the great names of nuclear physics. Historians have been fascinated by the physicists and have essentially adopted their point of view.

But what, after fifty years of basic research, made it possible to move from the first experimental reactor built in December 1942 to the bombs that exploded above the Japanese cities less than three years later? What accounts for the success in getting tens of thousands of people to work together, of stamping giant factories out of the ground in a few months, of mastering new technologies in rapid order? The Manhattan Project was not only a matter of cutting-edge research in nuclear physics. It also posed a set of technical problems. It was an industrial program, and the necessary know-how did not appear out of nothing; it had been forged over a half-century of learning the techniques of mass production in the high-pressure chemical industry, particularly at DuPont.

DuPont's participation in the Manhattan Project is known to the experts in the field. The official histories of the project note that DuPont's chemical engineers were in charge of the production of plutonium, the main component of the of the Nagasaki bomb.[1] David Hounshell and John K. Smith devote a few pages to the Manhattan Project's industrial aspect in their 1988 book *Science and Corporate Strategy: Du Pont R&D, 1902–1980*, but the role played by DuPont during the war is not central to their study, which on the contrary stresses the relation of science and technology to the market and to DuPont's commercial strategy.[2] For these two authors, as for Alfred Chandler, everything political—and what is more political than a war?—somehow lies outside the subject.

I do not intend here to narrate the details of a history simply because this has not yet been done. Some of the technical data I shall examine are known to specialists, others are not—in particular, as far as certain aspects of the DuPont organization and its links to the other partners of the project are concerned—but my objective here is very different from writing a new general history of the Manhattan project. What I have set out to analyze is the manner in which DuPont's chemical engineers were able to impose their way of doing things and their organization on their military and scientific partners. This will offer a new perspective on the project, one that is not meant to invalidate the existing ones but to bring out their hitherto neglected dimensions. This was also done by Thomas Hughes, who establishes a connection between the Tennessee Valley Authority and the Manhattan Project, both being joint systems of science and large-scale technology operating under the aegis of the government.[3] The difficulty is that the industrialists are placed in the background, as if they had been no more than secondary players in the service of the commissioning agency. As we shall see, things went very differently.

I have made use of the recently opened DuPont archives (including Crawford Greenewalt's notebooks), as well as of memoirs by Arthur Compton and General Leslie Groves and official histories that provide a detailed chronological outline.[4] I propose three sequences of analysis, that is to say, three ways of considering the Manhattan Project from the point of view of the role played by DuPont: the firm's integration into the project (first series); the building of the pilot plant (in which the chemical engineers imposed their culture, second series); the building of the plutonium production plant (the institutionalization of the chemical engineers, third series).

DuPont's Integration into the Manhattan Project

The Genesis of the Project

How did DuPont come to play a central role in the Manhattan Project? This question is not given enough attention by the official historian of the Manhattan Project, Richard Hewlett, who considered DuPont's participation "natural." But the fact is that the integration of the chemical company was fraught with a number of organizational and political problems. In order to understand this, we must briefly review the project's genesis and its organization.

At the very beginning of World War II, the idea that an atomic bomb might be built some day was already well established. In a letter of 2 August 1939 that has come down to posterity, Albert Einstein wrote to President Roosevelt, at the request of the physicist Leo Szilard, informing him that "it may become possible to set up a nuclear chain reaction . . . in the immediate future extremely powerful bombs of a new type may be constructed." The letter also mentioned the Germans' interest in this matter, adding, "you may think it desirable to have some permanent contact maintained between the administration and the group of physicists working on chain reactions in America."[5] This approach was backed up with a call on the president by Alexander Sachs, a Wall Street financier and informal advisor to Roosevelt.[6]

In October 1939, after some hesitation, Roosevelt decided to follow the recommendations of Einstein, Szilard, and Sachs. Cannon were already roaring in Europe, Poland had fallen, and it was known that German physicists were working on chain reactions and showing a keen interest in Belgian uranium as well as Norwegian heavy water.[7] The situation was thus sufficiently worrisome for Roosevelt to decide to create a uranium commission composed of high-ranking officers and physicists.[8]

From the first months of 1940 to the spring of 1942, research proceeded under the auspices of the National Defense Research Committee (NDRC), which had been established on 15 June 1940 to direct and coordinate scientific research with military objectives. It was headed by Vannevar Bush, a central figure of the American academic establishment at the time, professor of electrical engineering at MIT and president of the Carnegie Institution. The uranium commission was then a division of the NDRC. At the end of June 1941, the Office of Scientific Research and Development (OSRD) was created and given wider competencies than the NDRC. Bush became its president, while the chemist James Conant, president of Harvard, took over the direction of the NDRC, to be overseen henceforth by the OSRD. The uranium commission was thus part of two research groups until December 1941, when it was placed under the direct supervision of Bush and—for security reasons—given the anonymous name OSRD-1 Section.

At the time of the Japanese attack on Pearl Harbor and the U.S. entry into war against Japan and then Germany, the researchers were increasingly convinced of the feasibility of a chain reaction.[9] The projects were becoming more refined, particularly at the University of California, Berkeley, where Ernest Lawrence was working; at the University of Chicago, with the physicist Arthur Compton and his team; and at New York's Columbia University, where Enrico Fermi and Harold Urey were working.

Since the early 1930s, physicists and chemists had known, thanks to the work of James Chadwick, that the nucleus of an atom consists of protons and neutrons, and that electrons gravitate around this nucleus. The "atomic weight" of an atom is equal to the weight of its nucleus—the mass of the electrons being negligible. The atomic weight of the lightest element, hydrogen, is 1. At the other end of the periodic table is uranium, with a mass of 238, or 92 protons and 146 neutrons. The scientists also knew that certain elements exist in several forms, which differ only in the number of their neutrons. These forms are called isotopes. In 1934, the Italian physicist Enrico Fermi had demonstrated that the nuclei of atoms break up into neutrons and protons when bombarded with neutrons. Thereupon three German scientists, Lise Meitner, Otto Hahn, and Fritz Strassmann, discovered that by bombarding a uranium atom, one obtained two or three neutrons and a considerable release of energy.[10] Scientists thus believed even before the war that a chain reaction could be triggered with uranium, and that it would continue for as long as uranium atoms were present.

Natural uranium is composed of three isotopes: ^{238}uranium (99.3%), ^{235}uranium (0.7%), and a trace of ^{234}uranium. Experiments showed that ^{238}uranium captures too many neutrons and does not release enough in exchange (which is why natural uranium is not spontaneously fissile). By contrast, as the Danish physicist Niels Bohr demonstrated, ^{235}uranium could be split by rapid and slow neutrons and in return released more neutrons. This meant that only ^{235}uranium could be used to trigger a chain reaction. It also appeared that only a critical mass of ^{235}uranium could produce an instant chain reaction and thus serve as an explosive. The problem, then, was to separate the ^{235}uranium from the ^{238}uranium, even though these two isotopes are chemically identical.

Several methods exploiting the different atomic masses of these two isotopes were explored on the eve of the conflict, and all of them were put into operation during the war. Electromagnetic separation consisted of using a spectrometer to project uranium atoms into a magnetic field, and in this operation, the electrically most highly charged isotopes described a slightly different trajectory. At that point, it was possible to isolate them. Another method, called gaseous separation, consisted of passing the gasified uranium through a series of filters; and since the smallest elements passed through more quickly, they could be removed separately before they were allowed to solidify. A third means was that of thermic diffusion. Here the liquefied uranium is immersed in a cylinder traversed by hot steam. The ^{235}uranium attaches itself to this steam and follows it, whereas the ^{238}uranium remains stuck to the inner walls of the cylinder after it has cooled. The level of enrichment was low, but the equipment had the advantage that it did not take long to build. The Americans later used it as a means of pre-enrichment. A last possibility was centrifugation, which consisted of placing uranium in its liquefied state in a rotating cylinder. Here the ^{235}uranium, the heavier element, is propelled toward the outside by the centrifugal force. But this procedure involved too many technical problems and was therefore soon set aside.

Another option was entirely different from the methods of isotopic separation and consisted of exploiting the fact that ^{238}uranium does not absorb slow neutrons. If one could slow down, or "moderate" the stream of neutrons so much that it would only be absorbed by the ^{235}uranium, it would not even be necessary to separate the two isotopes. Fermi and Szilard felt that graphite was the most promising moderating element (heavy water, or deuterium, being another possibility) for bringing about this chain reaction. In the course of the year 1941, a group of physicists, including the young Glenn Seaborg, working under the direction of Ernest Lawrence at Berkeley, discovered that the capture of slow neutrons by uranium transformed this isotope into a new element, which they named plutonium. It had the same fissile qualities as ^{235}uranium.

No one knew which of these methods would be the most effective. The OSRD therefore decided to pursue all four of them simultaneously. Nor did anyone know how to build a detonating device, once ^{235}uranium or plutonium had been obtained. Everything was still extremely hazy, but everyone was agreed on the main objective, which was to produce enough fissile matter.

In the spring of 1942, Bush and Conant were already envisaging the semi-industrial and industrial stages by which plants producing fissile materials would be built. In May, they recommended that the S–1 committee apply to the Army, and more specifically to the Corps of Engineers, which had the competencies for undertaking large-scale building projects.[11] The Army Corps of Engineers was organized by geographical divisions, which in turn were divided into districts. The nuclear project was set up as one district, which by way of exception was directly answerable to the Chief of Engineers: this was the Manhattan Engineer District. Colonel James C. Marshall was its first head, but he was soon replaced—on 17 September 1942—by Colonel Leslie Groves, who was promoted to brigadier general a few days later. Groves, who was forty-six at the time, had occupied various administrative positions in the Corps and had notably been involved in the building of the Pentagon.

By the middle of 1942, the operations were divided into two major categories: scientific research under the supervision of the S-1 committee; and the construction of the Manhattan District project. The military engineers asked the engineering firms Stone & Webster and M. W. Kellogg to study the construction and engineering problems of the different projected plants. The plan called for grouping them together in Tennessee near the hydroelectric installations of the Tennessee Valley Authority. Stone & Webster would be in charge of the plants for the centrifugal and electromagnetic process and the plutonium production; Kellogg would be responsible for the gaseous diffusion plant.[12]

This was the situation in the autumn of 1942, when DuPont came into the picture. On 30 October 1942, Willis Harrington, vice president at DuPont and a member of the Executive Committee, received a telephone call from General Groves asking him to come to see him "to discuss a matter of great military importance to the United States."[13] The project had turned out to be too big for Stone & Webster, which had to honor a large number of other military commissions and would have to subcontract some of the work even if it became involved. On 26 September 1942, Groves and the directors of Stone & Webster had therefore agreed to ask DuPont to provide assistance in building the plutonium plant. But at this point, Groves and Compton already felt that the scope of the task would make it necessary for DuPont to take over the operations in their entirety, and their opinion was endorsed by Bush and Conant in early October.[14]

Moreover, entrusting all of the operations to a single firm could be expected to make for increased efficiency. DuPont was the only major chemical company that built its own plants and equipment, thanks to its Department of Engineering, and this was well known to Groves, himself an expert in the field.

But when Compton suggested to his physicist colleagues in Chicago that Du-Pont might become involved in the project, there was a "near-rebellion," as he recalled later.[15] Many physicists felt that they were capable of handling the tasks of development and construction on their own. Moreover, they considered DuPont to be the most repulsive kind of big business, a "merchant of death" company whose political reputation smelled to high heaven. Greenewalt recalls that certain physicists thought that DuPont was looking for profits from nuclear energy after the war:

> The trouble was almost wholly with the physicists . . . they thought we were there in our own interest, that we were (this quote was actually used) "picking their brains" with the idea that we would make a barrel of money out of atomic energy. I would say, "not at all, we're not in the power business." Power was the only thing we could see [in it] and we bought our power plants. I made a bet with one of the fellows, Herb Anderson [an assistant of Fermi's], I said, "I'll bet you five dollars that three years after this is over, we will be out of atomic energy completely." I won the bet, but I never got the five dollars. . . .
>
> That didn't last long because I think they saw immediately that we were really in there to do a job.[16]

By July 1942, some DuPont employees were already working at the "metallurgical laboratory" of the University of Chicago (the code name for Compton's group of nuclear physicists). One of them was Charles Cooper, a chemical engineer and associate of Thomas Chilton's, who had come to Chicago on 30 July and was joined between July and October by eight other DuPont employees. Cooper had no doubt been asked to participate on the recommendation of his former professor at MIT, Warren K. Lewis, a friend of Compton's.[17] Until October, DuPont was simply making some of its employees available to the project; the company as such was not engaged. Throughout the month of October, negotiations took place among Groves, the Chicago physicists, and DuPont's Engineering Department (specifically the chief engineer, Everett Ackart, his assistant Thomas Gary, and Thomas Chilton). The discussions focused on the help Stone & Webster would need for the construction of the pilot plant that was projected to be built in the Argonne Forest, not far from Chicago.[18] DuPont's Executive Committee was, of course, kept informed of these approaches, and on 7 October, it authorized DuPont's participation in the design and the supply of materials for the pilot plant.[19]

The day after the telephone call from Groves, Harrington and Charles

Stine, another vice president of DuPont and former head of the Department of Chemical Research, went to Washington to meet with Groves and Conant, who explained everything to them: the project, the status of the work, the need to proceed to large-scale production, and their "grave doubts about the feasibility of the undertaking."[20] They requested that DuPont give an opinion about the project on the basis of the scientific data to be furnished by the University of Chicago. Harrington and Stine pointed out that DuPont had no experience with nuclear physics, and that they felt incompetent in this field. To which Groves replied that "there was no one qualified, that he needed advice badly, and that in view of Du Pont's broad industrial experience, he preferred to 'hang his hat' on Du Pont's opinion and judgment than to have no opinion at all."[21] He candidly added that theirs was the only company capable of designing, building, and operating such a plant for large-scale production. Harrington and Stine responded that they would report on the matter to the firm's Executive Committee.

DuPont's Executive Committee thereupon decided to send to Chicago an evaluating committee composed of Stine, Roger Williams (director of the Ammonia Department), Crawford Greenewalt (director of the Grasselli Department of Heavy Chemistry since 1940), Chilton, Gary, C. R. Johnson, and Pardee of the Department of Engineering. General Groves also went to Wilmington to meet with DuPont's Executive Committee and its president, Walter Carpenter, who had replaced Lammot du Pont in 1940. The report of the committee that had been sent to Chicago and the review of Groves's visit betray great wariness of the project on DuPont's part. Nonetheless, the firm's representatives added at every turn that these were critical days, that the work demanded of them was "of extreme importance to the country," and that "the enemies of the United States in this war were working to produce plutonium or its equivalent."[22] In order to gain the clearest possible understanding of the industrial task to be undertaken, and also in order to involve DuPont, Groves and Conant had the idea of setting up a five-man evaluation committee headed by MIT's Warren K. Lewis, whose other members were Williams, Greenewalt, and Gary from DuPont and Eger Murphree of Standard Oil (although the latter fell ill and did not attend the meetings).[23] Williams, Greenewalt, and Murphree had been Lewis's students at MIT, and the five men knew one another well.[24]

Another connection deserves to be noted here as well: that between the Manhattan Project and Catalytic Research Associates, a joint venture of Standard Oil, M. W. Kellogg, and IG Farben that had been created in 1938 for the purpose of developing a new continuous-flow process of catalytic cracking. Bringing together several hundred chemists and engineers, it represented, according to Thomas Hughes, the greatest concentration of scientific and technical capital before the Manhattan Project.[25] We know practically nothing about the manner in which techniques coming out of petrochemistry also structured the Manhattan Project, but it may be similar to what happened in the case of DuPont.

The evaluating committee visited Columbia, Berkeley, and Chicago, where on 2 December 1942, Greenewalt witnessed the first chain reaction in history: in a squash court under the football stadium of the University of Chicago, requisitioned for the occasion, Fermi and his team had placed uranium and graphite rods into a reactor and triggered a continuous reaction that emitted heat and neutron radiation. When the neutrons struck uranium atoms, these split and released more neutrons at an exponential rate. When the reactor began to heat up dangerously and the radioactivity increased, bars of cadmium (which absorbs the neutrons) inserted into the reactor smothered the reaction.[26]

This achievement clearly had an effect on the report of the evaluating committee, for it expressed the opinion that Fermi's nuclear reactor was most likely to be successful. However, it added that the isotopic separation processes should be pursued as well.[27] It is probable that the DuPont directors were impressed by Fermi's achievement and that this contributed to lifting their last hesitations. Matters came to a head in the last weeks of the year 1942. The period of negotiations and discussions ended on 21 December, when a contract was signed between the U.S. Army and DuPont, stipulating that DuPont was in charge of building and operating the future plutonium plant.[28]

Why DuPont?

Initially, this decision by DuPont's directors was motivated by factors unrelated to the company and linked to the general context of the war and the state of the Manhattan Project in late 1942. At that point, the outcome of the war was highly uncertain: the Germans were embarking on their second winter campaign in Russia, and the Wehrmacht hoped to end it with a last thrust in the region of Stalingrad. Rommel's Afrika Korps was advancing toward Egypt, and the Japanese, weakened though they were by the battle of Midway, nonetheless were still a great threat and remained ensconced in many places throughout Asia and the Pacific islands. It was a critical point, and it was difficult for anyone to ignore the pressure of patriotism. DuPont could not very well elude the urgent requests of the military.

Then, too, the Manhattan Project had reached a critical turning point. The first stage, involving small committees, research teams dispersed among several universities, and the discreet role of the Army and some engineering firms, had come to its end with the first chain reaction. The experimentation of December 1942 signaled the end of trial and error. The time had come to enter an industrial phase, which meant bringing in the major companies, since they commanded the necessary know-how. Thus DuPont was called upon to work on plutonium, Union Carbide on gaseous separation, and Eastman Kodak on electromagnetic separation. The engineering firms faded into the background and were reduced to the role of minor players. The Lewis Committee rightly recommended that "companies with appropriate experience should be entrusted with the necessary responsibilities for carrying out these projects."[29]

This was also the moment when General Groves took command and ordered the research activities for the bomb to be grouped together at Los Alamos, a desert site in New Mexico; the building of plants at Oak Ridge, Tennessee, and later at Hanford, Washington; and research activities for fissile materials to continue at Chicago, Berkeley, and Columbia. In short, the Manhattan Project assumed a different scope in late 1942.

It is also important to consider factors internal to the company. The hesitations of DuPont's directors are perceptible when one analyzes the contract negotiations that took place in December 1942. DuPont actually refused any profits and agreed to sign the contract only on this condition.[30] The company would be reimbursed for its expenses and receive one symbolic dollar a year. (The cost-plus-fixed-fee contract, which protected businesses from runaway expenses, had been instituted by Congress in December 1941 to expedite the mobilization of American industry. Beyond that, it prevented the reaping of profits prorated to the expenses incurred, thereby avoiding polemics of the kind aroused by DuPont's profits during World War I.) The aim was to defuse any future charges that DuPont was a "merchant of death," and nuclear death at that: the directors had been told that the projected bomb would be unbelievably powerful, that the enemy would be unable to repair the damage it would cause, and that its manufacture itself would not be free of major risks for the employees and for the population of the surrounding area. DuPont had very bad memories of the Nye Commission and did not want its promising postwar activities to be blighted by accusations of morally reprehensible profits.

DuPont's prudent lawyers nonetheless pointed out that if the company went over budget, the government would have to pay and that, financially, the operation should be a good deal.[31] It was stipulated, moreover, that the contract had been drawn up at the express request of the Army, and that it would expire immediately upon the cessation of hostilities. DuPont also requested that the president of the United States give his formal consent to the enterprise. This request was forwarded in a letter of 16 December from Vannevar Bush to President Roosevelt.[32] As an extra precaution, a "secret letter" signed by the government, the University of Chicago, and DuPont laid out in clear language the state of knowledge, the reasons why the building of an atomic bomb had been decided on, and the possible dangers involved. Clearly, DuPont meant to erect a legal rampart that would protect it from any future congressional investigation of its nuclear activities.[33] The final contract was signed by the Executive Committee on 6 October 1943, to be effective retroactively.

Going into the nuclear field meant not only a change in the scope of financial investments—which was nothing to faze the directors and engineers of a company accustomed to costly projects—it also meant a cultural change for Du Pont, in the sense that the Manhattan Project required the establishment of close and simultaneous relations with other scientists and with the government. Did DuPont's directors realize this from the start? They had no way of knowing in advance either how the operations would be organized or, especially, that their firm would become more and more deeply involved in this project. But they already suspected that this project would be altogether different from the usual military commissions, that it would be unique, not only in scope but also because of the nature of the relations that DuPont would have to establish with scientific and military partners. Stine in particular, a close associate of the du Ponts, feared the "political" aspects of the project. He was right. The Manhattan Project was eminently political. As the members of the evaluation committee had realized, nothing had jelled with respect to the scientific and technical options and the power relations among the different partners. Unlike DuPont, with its almost military organization of the projects it undertook, Manhattan left all participants in a gray zone as to their margin of maneuver. Such initial indecision is probably characteristic of all large-scale projects, but nothing in the history of the firm had ever been so indeterminate. DuPont had never been involved in so hypothetical an adventure before.

To be sure, DuPont was accustomed to dealing with the military, which had always been a major customer, and the company had long had several devoted operatives to represent it in the corridors of the War Department. But until now these relations had been based on commercial contracts for supplying gunpowder and explosives. In that scenario, DuPont delivered and the government paid. And indeed, this is what happened during World War II as far as conventional explosives and smokeless powder were concerned: just as in the Great War, DuPont garnered huge contracts and made fabulous profits. The company produced 70 percent of all the explosives manufactured in the United States during the war, or 2.3 million tons—which, incidentally, is three times as much as was produced during World War I. This figure included, among other products, 700,000 tons of TNT and 1.5 million tons of smokeless powder—or two and a half times the 1914–18 total.

But unlike in the previous conflict, when 80 percent of DuPont's production had consisted of explosives and gunpowder, this time these products accounted for only 25 percent.[34] DuPont was also making paint for the Navy, dyes for uniforms, antifreeze for airplanes and tanks, cellophane for wrapping soldiers' rations, synthetic rubber for the tires of trucks and bombers, insecticides like DDT, waterproofing agents, and nylon for parachutes. By contrast with metal-producing industries, which usually had to reconfigure their entire production sites, the chemical industry was able to devote itself to the war effort immediately and without major investments. By the same token, the return to civilian production would not take long. This was a case of maximal synergy between the military and the civilian spheres.

The production of nylon steadily increased throughout the war, served by a second plant that was built near Richmond, Virginia. But it was not enough, and the government asked American women to donate their nylons to the war effort. To encourage them, Hollywood actresses were approached and gladly agreed to take off their nylons in front of the cameras.

All the transactions between DuPont and the government were conducted on a strictly commercial basis. As Lammot du Pont clearly put it at a meeting of the National Association of Manufacturers in September 1942: "Do business with the government as you would with any other buyer. If it wants to buy, it has to do so at your price."[35]

However, in the context of the Manhattan Project, the government was no longer exactly just another big client, but rather a partner in the problem-laden manufacture of a product that so far existed only as an idea, with a final objective that was highly uncertain. The success of Fermi's experimentation on 2 December 1942 notwithstanding, there was still much doubt as to whether plutonium could be produced on an industrial scale, particularly in light of pessimistic information about it just then arriving from Great Britain.[36] This undertaking definitely did not appeal to DuPont's directors, but they had little choice. Becoming involved in the Manhattan Project required the establishment of a complex system of relationships among private and public partners, and this was difficult to conceive for a company that saw its relation with the federal government and its bureaucracy of experts in terms of strict opposition. A strong desire for independence, combined with the political conservatism of its directors, did nothing to further DuPont's integration into the multipolar public-private network of the Manhattan Project. In the end, one other factor may have played a role: the pride of the company and the DuPont men. Vannevar Bush, James Conant, General Groves, and the president himself, who kept close tabs on this project, came knocking on DuPont's door, arguing that the Manhattan Project might well decide the outcome of the war and tip the scales for the good side, and that DuPont had a crucial role to play for the higher interest of the country and the free world. For men whose skins had been toughened by long-standing criticism and who lived with a persecution complex, being courted in this manner was flattering indeed. Interviews with some of the engineers and the company's internal *Du Pont Magazine*, which after the war published articles on the manufacture of plutonium, betray pride in the company and a quiet certainty that they were essential actors without whom nothing would have been accomplished.[37] Hence DuPont was somewhat resentful of the physicists, whom it implicitly accused after the war of having "pulled the covers" to their side and influenced the official histories. Greenewalt confided with some bitterness that "none of the people that have written histories have ever come to me—Leona Marshall Libby, Arthur Compton, Groves, or the official historians."[38] Later he added:

> it is unfortunate that the histories that have been written about this thing all come out with the Hanford development as a triumph for the physicists. That's the way it always comes out; it was the physicists' thing. For example, the physicists never gave any thought to the chemistry at all. I said at this meeting that to my way of thinking, it was one of the greatest interdisciplinary efforts ever mounted. . . . Oppenheimer, who was a great guy, and of whom I was really very fond, a charming person, but he had his touch of arrogance too. He made this statement, "the physicists have known sin." I said, "my God, if everybody that has made an important contribution to the Hanford project was a sinner, then it would take several Rose Bowls to hold them all." I considered the thing an interdisciplinary effort . . . chemists, engineers, all sorts of people from all over the

country putting their back in and getting this darn thing done. I would say that the accomplishment of the chemists was just as great as that of the physicists.[39]

DuPont's directors and engineers were torn between wariness of a risky and potentially damaging project and the pride of knowing that they were indispensable. In the end, the collective memory of the company has come to reflect this ambiguity. Depending on the occasion, it will either recall the feats it accomplished—as it did when General Groves's memoirs, *Now It Can Be Told,* came out in 1962—or exercise a certain discretion concerning an industrial epic that ended with the total destruction of Hiroshima and Nagasaki and unspeakable suffering.

The decision-making process at DuPont can be analyzed in a cultural perspective that treats the company as the site of social representations that had created identities and constraints over the preceding decades that militated against engagement in the Manhattan Project, and that were only overcome by very strong outside pressure and special circumstances. Neither the pressures of the market nor the strategic choices of the directors were able to modify DuPont's representations and structures: powerful political pressure was needed to goad the firm to set out in a new and unexpected direction. To object that the war was an atypical parenthesis and that the culture of DuPont reasserted itself as soon as the fighting was over would, moreover, be to ignore the ways in which the technical and interrelational practices of the Manhattan Project transformed the company.

The Pilot Plant

The first stage at DuPont consisted of getting organized. The Executive Committee decided that the responsibility and the authority for the work would be formally given to the Department of Explosives. On 16 December 1942, a new division was created, the TNX division of the Department of Explosives. Why this name, TNX? Like most of the designations in the Manhattan Project, TNX was a code name and referred falsely to trinitroxylene (TNX), a powerful explosive derived from TNT that had been developed during World War I.[40] In reality, of course, this was not a matter of producing a conventional explosive, and in practice this TNX division had the status of an autonomous department reporting directly to the Executive Committee.

DuPont's organizational capacity, characterized by a decentralized, multidivisional structure—which was increasingly adopted throughout American industry—turned out to be largely adequate for a collaboration with military and civilian partners. This was, first of all, because this decentralized structure allowed for the rapid addition of a department or a division when a new production line was to be started. Secondly, TNX, like the other divisions, was autonomous and had its own organization, which allowed it to function with extreme discretion, unbeknownst to all but those directly involved. The managers and the administrative services of TNX worked at company headquarters in Wilmington. Their offices were under constant surveillance of guards, who checked the passes of all comers.[41]

TNX's assistant general manager was appointed immediately, on that same 16 December. He was Roger Williams, until then the extremely enterprising director of the Ammonia Department. TNX was divided into Technical and Manufacturing Divisions (fig. 4.1).

The Technical Division had "the prime responsibility of maintaining liaison with Chicago to procure and interpret the technical data necessary for the work."[42] The manager of this division was Crawford Greenewalt, whose career has been outlined above, and who was expected to spend half of his time in Chicago with Compton's physicists. This division counted only a half-dozen engineers, sometimes a few more.[43] Greenewalt's assistant was George Graves, a chemical engineer who had also come from Nylon. Also in the Technical Division were Lombard Squires (Ammonia), in charge of chemical separation; Hood Worthington (Nylon), responsible for the engineering of the reactor; and Dale Babcock (Nylon), for the physics of the reactor and for safety.[44]

The Manufacturing Division was responsible for the various aspects of designing and building the equipment (plant, reactors, etc). It worked closely with the Technical Division, as well as with DuPont's Department of Engineering, which was also concerned with matters of design and construction and had charge of recruiting and training personnel. Its manager was R. Monte ("Monty") Evans, a civil engineer who like Williams came from the Ammonia Department.

On 1 February 1943, TNX counted 44 employees; by 1 March, it had 90. These figures do not include the personnel of the Department of Engineering assigned to the Manhattan Project, but it appears that 90 percent of that department's personnel and resources were devoted to the project.[45] In addition, a total of 392 DuPont employees were lent to the University of Chicago and its "metallurgical laboratory."[46]

It is difficult to find out exactly how Williams and Greenewalt were chosen.

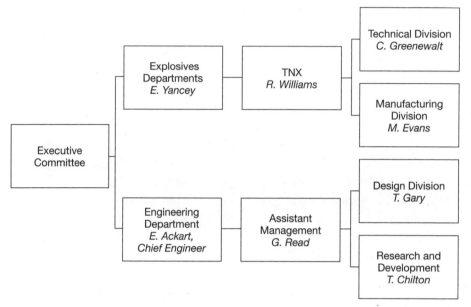

Fig. 4.1. TNX and the Engineering Department at DuPont

But it does seem clear that DuPont's Executive Committee had chosen them as early as autumn 1942, that is to say, even before the company had become formally involved in the project. Significantly, the group of engineers that visited the metallurgical laboratory at Chicago on 4 November 1942 was made up of Williams, Greenewalt, Gary, Chilton, Stine, and Bolton. Greenewalt had "no idea" why he had been chosen. "They might have picked Roger [Williams] just the way they picked me; two people whose background had been research, and research with a predominance of engineering problems. That might have been it. Roger was considerably older than I and had more experience."[47] The Executive Committee chose its elite engineers, men who had proven their mettle in the development of nylon a few years earlier, and who had the scientific background to engage in dialogue with the physicists at Chicago. Williams had extensive managerial experience, whereas Greenewalt was principally a research engineer and had only recently been promoted to the position of director of research in the Grasselli Department of Heavy Chemistry. Greenewalt and Williams then picked their men, engineers whom they knew well because they had worked with them.

Although placed, in principle, under the authority of the general director of the Explosives Department, Ted Yancey, the technical culture of gunpowder and explosives was irrelevant to TNX: the engineers who worked in that division came out of the new technical culture examined above, a culture that had crystallized during the development of nylon a few years earlier. As Greenewalt explained quite simply: "The people who were making dynamite and smokeless powder had not the remotest idea of what we were doing."[48] The powder and explosives people were the direct heirs of the old chemical firm, where experience and practical know-how predominated. They were quite different from Greenewalt, Williams, and their collaborators, research engineers with academic degrees who favored a more theoretical approach. A remark of Greenewalt's ("it was not their line") betrays a certain condescension toward the home-grown engineers of the Explosives Department. Greenewalt had a precise knowledge of degrees and careers. When Hounshell mentioned that Gary was an MIT alumnus, Greenewalt corrected him: "'As a matter of fact, I would doubt very much if it was true in Tom Gary's case, I don't think he went to MIT [which was the case]."[49]

Williams, Greenewalt, Cooper, Graves, and the others, having worked in the same departments, shared the same way of approaching problems of production, the same inclination to establish a formal scientific methodology. They were self-assured, proud of their ability to bring large-scale projects to a satisfactory conclusion, and by and large aware of being the best engineers in the company. When Greenewalt said that he did not know why the DuPont directors had chosen him, this does not mean that he was surprised by this appointment, but rather that he did not know the precise circumstances that had led to it. The encounter between these engineers and the physicists, then, brought together two communities that had different practices and different cultures and little interest in compromise.

The Building of the Test Reactor

The formal mathematical methodology of chemical engineering was a good preparation for the exigencies of nuclear physics. In the 1930s, the chemical engineers of the Department of Chemical Research had done a great deal of research on heat transfers and fluid dynamics. The work done at that time under the direction of Thomas Chilton had essentially been aimed at meeting the needs of the Ammonia Department with its high-pressure equipment, and later those connected with the development of nylon and polymer chemistry. Eventually, this work also had its usefulness for the Manhattan Project.

Nuclear engineering, that is, the technological know-how needed for the construction of atomic facilities was born with the Manhattan Project and delivered by DuPont's chemical engineers. It can be roughly divided into two major fields. The first concerns nuclear physics in the narrow sense, namely, the study of nuclear reactions—neutron interactions and gamma radiation. The engineers must determine the number of neutrons at a moment t in relation to different data (temperature, mass of graphite and uranium), as well as the best ratio k possible (k being the ratio that measures the acceleration of the chain reaction.)[50] All of this was not known at the time, and the engineers proceeded by trial and error. Greenewalt's personal diary (some parts of which are still classified as "defense secrets") records the hesitations, the day-to-day advances, and the experimentation involved.[51] The engineers knew nothing about nuclear physics. Hence the importance of TNX's Technical Division (headed by Greenewalt), which served as liaison between the physicists and the engineers of the Manufacturing Division and the Department of Engineering. The equations of the physicists had to be translated into engineering language. This translation work was essential to the project.

The second area of nuclear engineering deals with more conventional physical and physico-chemical problems—heat transfers, thermodynamics, chemical separation. This last area was not unknown to the DuPont engineers. Thanks to their experience with high-pressure chemistry, it was relatively familiar territory for them, even if radioactivity, the corrosive nature of some substances, and the extreme toxicity of plutonium gave rise to novel security concerns, calling for containment or manipulation at a distance. A final difficulty had to do with the absolute precision that was called for. As Thomas C. Wilson, a construction engineer, recalled, "the building constraints involved two changes from the way I used to do things: the lack of skilled labor and the precision of the work, which was fantastic. I remember one time when we worked 72 hours at one stretch to make sure that one large building was absolutely, and I mean absolutely, airtight."[52]

TNX, then, was expected to produce plutonium in quantities sufficient to make bombs. This assignment involved two major tasks: one was to build a pilot plant at Oak Ridge, Tennessee; the other to build the production plant at Hanford, Washington State.

The objective of the pilot plant was to obtain a series of data before proceeding to large-scale production, to produce a small quantity of plutonium for research and testing purposes, and also to train the DuPont personnel. Yet the December

1942 contract mentioned only the construction and operation of the production plant, assigning the responsibility for the pilot plant to Stone & Webster and the University of Chicago. But once again the task fell to DuPont, and an addendum was included in the contract.[53] The Executive Committee voted against it, but eventually complied. The sibylline and convoluted resolution of 15 January 1943 is indicative of DuPont's reticent commitment to this task:

> After full discussion, it was moved and carried (Vice-presidents Jasper E. Crane and W. F. Harrington voting in the negative) that the Explosives Department be authorized to inform the representatives of the Government that, due to our complete lack of knowledge and experience with processes of this nature, we feel that the best interests of this development are not served by the Du Pont Company assuming the responsibility for the operation of the projected experimental plant to be erected in Tennessee, until those investigators, now assigned to the development of this process and who have brought it to its present stage of development, have demonstrated by operating that process in the experimental plant that it can be operated on the experimental plant scale.
>
> However, it is further the sense of the Executive Committee that the representatives of the US government be advised that the Du Pont Company is prepared to assist the Government and its technicians by supplying such advice and manpower as it is able to in connection with the administrative and operating problems involved in the operation of this experimental plant.[54]

The question of how the work was to be divided between the DuPont engineers and the Chicago physicists was an issue from the very beginning of their partnership, and several conflicts erupted over this during 1943. This was, first of all, because DuPont's appearance on the nuclear scene was very upsetting to many of the scientists. Then, too, there was a series of controversies that pitted the two parties against each other. The first of these concerned the siting of the pilot plant. The Chicago scientists wanted to place it close to Chicago, at Argonne. Fermi also wanted to keep the test reactor—the one that had served for the first chain reaction on 2 December 1942—on the Chicago campus.[55] DuPont and its engineers pointed out that a nuclear reactor on a university campus was too dangerous, and furthermore that the site in the Argonne forest was too small for the pilot plant. A compromise was reached, and Fermi's reactor moved to Argonne, while the pilot plant was established in Tennessee. But it took several weeks more to decide who should run the plant, DuPont or the University of Chicago. In the end, in line with DuPont's wishes, the university undertook to do so, this arrangement having been approved by its president, Robert Maynard Hutchins. But the wrangling left traces, as Compton recalled in his 1956 memoir *Atomic Quest: A Personal Narrative.*[56] Williams had to promise that DuPont would be entirely responsible for the construction and the proper functioning of the pilot plant. But this was to reckon without DuPont's Executive Committee, which refused to take on this project, arguing that the pilot plant should be operated by scientists rather than by engineers. James Conant, for his part, felt that the initial agreement had to be honored. A compromise was finally reached: DuPont built the pilot plant at Oak Ridge, but its facilities were operated by the physicists—who were quite satisfied with the arrangement.[57]

Ground was broken on 22 February 1943, under Greenewalt's supervision, at an isolated spot near the small town of Clinton, in an area called Oak Ridge. The site was some twenty kilometers from Knoxville and some of the Tennessee Valley Authority hydroelectric facilities.[58] Construction of the separation plant began on 9 March, and of the reactor on 23 March, and the work progressed throughout 1943. One notable difficulty, the recruitment of labor, was a general problem during the war, especially in regions on the periphery of major centers of employment. Between 1 February 1943 and 1 February 1944, 6,041 workers and technicians were employed, but many more were hired, given an average monthly turnover of 24 percent, sometimes rising as high as 35 percent.[59] The building project as a whole was supervised by TNX and the DuPont Department of Engineering, as well as by a team of Chicago physicists, among them the young John A. Wheeler, who had worked under Niels Bohr in Copenhagen in the 1930s. This team at first worked at the DuPont headquarters in Wilmington, but it was transferred to Tennessee in August 1943. Wheeler, however, stayed in Wilmington.[60]

Who was to design the equipment? Eugene Wigner, who was responsible for designing the water-cooled reactor, would have liked to be put in charge of the planning for the production plant as well. But Greenewalt pointed out that DuPont's usual practice was to place engineers in charge of production, although they could always ask the scientists for information.[61] This did not sit well with the scientists, who wished to keep general control of the project. They quickly realized that this was not what DuPont had in mind, and that in fact power had changed hands. With the arrival of DuPont, engineers and managers had gotten ahead of the scientists. This is what Greenewalt expressed in his own way when he kept saying that the physicists were not easy to work with. As he confided:

[Compton] had people like Szilard and Wigner, who suffered from a general disease that brilliant people, particularly in physics, seem to have; that is, because they're brilliant in their own field, they think they know everybody else's. Wigner would have had not the slightest hesitation in telling us how to run the Du Pont Company. As a matter of fact, all of the difficulty—and there was a great deal during that design—was their certainty that they knew better than we how to go at this problem. . . .

They built a chain reacting pile with their own lily-white hands. . . . All right, but nonetheless they said, "well the rest of it is just a matter of engineering" . . . as if [engineers] were glorified plumbers that had no science of their own.[62]

What was at stake here, to put it in psychologizing terms, was control of the project. The physicists were convinced that the DuPont bureaucracy would paralyze the project, that the engineers spent too much time on safety concerns, and that it would be DuPont's fault if nothing was ready in time. Hewlett notes that in late July 1943 several physicists—including, it seems, Wigner, working behind the scenes and with the consent of Fermi—were convinced that DuPont was sabotaging the project, and in July 1943, they sent a letter to Eleanor Roosevelt, who passed it on to her husband. In addition, the physicists contacted a close adviser to the president, Bernard Baruch, a major figure in the Democratic Party, who had been in charge of war production during World War I and was on excellent terms with the Roosevelts, as well as Supreme Court Justice Felix Frankfurter.[63] They wanted to obtain authorization to build a reactor operating with heavy water, which has the advantage of serving as both coolant and moderating element, whereas the engineers had already chosen graphite and water for the production reactors. [General] Groves calmed down the physicists by appointing an evaluation committee, which confirmed DuPont's choice but also recommended the building of a test reactor using heavy water at the Argonne site. A few million dollars no longer mattered, and the main thing was to avoid a complete break between the engineers and the physicists.

The pilot plant was composed of two units: a nuclear reactor consisting of 1,248 uranium rods inserted in approximately 500 tons of graphite for the production of irradiated uranium; and a chemical separation unit, where the plutonium would be extracted from the irradiated uranium. Hanford was to repeat this arrangement on a large scale. The main problem that had to be dealt with was the cooling of the reactor. Irradiating the uranium in the reactor generated considerable heat, which had to be reduced as much as possible without compromising the chain reaction.[64] The DuPont engineers and the Chicago physicists tried their best to choose an element that would cool the reactor and yet absorb a minimum of neutrons (so as not to smother the chain reaction). Helium and water were two possibilities, and the choice was made to explore them simultaneously.[65] For the time being, the engineers decided to cool the reactor with air, which could be done because of its small size.

It also became necessary to think about a containment wall that would prevent the contamination of the surrounding area. The Department of Engineering, in liaison with Chicago, studied the best possible means of protection. They decided on a concrete and steel wall two meters thick to enclose the reactor, which was pierced on two opposite sides by holes for inserting the uranium rods and pushing them out again after irradiation. Once the uranium had been irradiated long enough, a robot controlled from a distance pushed on the rods, which then fell into a basin of water, where they would remain for several weeks until their extremely high radioactivity decreased somewhat. Subsequently, they were transferred to the chemical separation plant in sealed crates.

As for the chemical separation, nobody in early 1943 was sure which process should be used. The irradiated uranium contained only a very small quantity of plutonium (less than 1%), which had to be isolated and extracted. The chemist Glenn Seaborg had determined that the best way to achieve this was to obtain a precipitate. In other words, the irradiated uranium had to be diluted in a solution containing a precipitant that would bind the plutonium and then separate the plutonium from the precipitant. This, unlike isotopic separation, was a purely chemical operation. In May, it was decided to adopt a process using bismuth phosphate.

In late November 1943, the reactor was started up. By early 1944, the first grams of plutonium had been shipped to Chicago and to Los Alamos, where the design of the bomb was being studied under the direction of the physicist J. Robert Oppenheimer.

Who Controls the Operations?

How, concretely, did the collaboration of Du Pont, the scientists, and the military function? The DuPont engineers were in charge of building the plant. To do this, they had to mobilize labor and find scarce materials, and if they encountered difficulties, they would contact Groves or his assistant Colonel Kenneth Nichols (or, for minor problems, one of the officers assigned to Wilmington or Clinton) to obtain the necessary authorizations and requisition orders. The Manhattan Project was classified "AA3" in the order of priority for the war industry. This meant that it came after "AA1" and "AA2," which generally applied to supplies of munitions and essential materials for tank and aircraft production. The AA3 rating was usually sufficient, but in urgent situations an "AAA" order (absolute priority) could be requested, and this was done frequently.

DuPont also stayed in close touch with the physicists of the "metallurgical laboratory" at the University of Chicago: some of the company's engineers, notably Greenewalt and Charles Cooper, were ensconced in Chicago and participated in the meetings of the physicists, while some of the physicists worked in Wilmington and at Oak Ridge. John Wheeler in particular became a permanent member of the DuPont team in Wilmington. There were constant exchanges of personnel among Chicago, Wilmington, Clinton, and, soon, Hanford. Greenewalt stayed at the Shoreland Hotel near the Chicago campus, spending half of the week there and the other half in Wilmington.[66]

Since the technical reports of the DuPont engineers had to be approved by the Chicago scientists, they were submitted for their signature. Disagreements arose frequently, and in such cases, the reports were shuttled back and forth until an agreement was reached. The DuPont engineers often complained that the physicists "hated to deal with the blueprints they were sent." This revealing statement indicates that physicists and engineers did not speak the same language. Engineers express themselves by means of blueprints, which visualize the product or process under consideration and "should be interpreted as a conventional tool that provides an underlying agreement on what is supposed to be produced."[67] The blueprint establishes a kind of collusion among engineers, welds together a technical community, and excludes those who do not speak its language. Physicists, on the other hand, express themselves by means of calculations, clusters of equations. Greenewalt recalls that when Sam Allison, the director of the Chicago laboratory, examined a blueprint, he "always managed to get in some sort of snide remark."[68] The reason was that a blueprint *did not mean anything* to a physicist like Allison. This is corroborated by Richard Feynman, the future Nobel Prize laureate in physics, who asked: "[H]ow do you look at a plant that isn't built yet? I don't know! . . . Lieutenant Zumwalt, who was always coming around with me because I had to have an escort everywhere, takes me into this room where there are these two engineers and a looong table covered with a stack of blueprints . . . I'm not good at reading blueprints."[69]

Greenewalt correctly observes that "since the two groups spoke two different languages, it was a task of interpretation, taking the research results and translating them into terms that could be used to design a piece of equipment."[70] But this translation work of his was not fair to the physicists when it came to the control of the operations, which was conferred precisely by a thorough understanding of the blueprint, its special language, and the changes it called for. Here the translator has a strategic position: the apparent modesty of his role in reality conceals a transfer of power from the "theoreticians" to the "practitioners" (fig. 4.2).

In addition, the blueprint had the effect of fixing the features of the planned construction once and for all, unlike scientific calculations, which can be modified ad infinitum. The calculation remains malleable, whereas the blueprint settles and congeals the technical choices. The DuPont engineers had to settle on the essential features of the plant as quickly as possible and begin the construction without necessarily having all the data; for they could not constantly add changes and correspondingly delay the beginning of the operations, although they did leave plenty of margin for contingencies. Hence the resentment of the physicists, who felt that they were being cut out of the project when it was no longer expressed in their language. Even the building of the pilot plant initially did not have their approval: Greenewalt had explained to them that DuPont always tested its projects, that a pilot plant had been built when nylon was being developed, and that this was an efficient way to detect technical problems at an early stage, but the scientists had only begrudgingly accepted this principle.[71] As Fermi privately told Greenewalt: What you should do is to build a pile just as quickly as you can, cut corners, do anything possible to get it done quickly. Then you will run it, and it won't work. Then you'll find out why it doesn't work and you will build another one that does.[72]

Fermi and his friends at Chicago were hoping to hold on as long as possible to their maneuvering room and to the possibilities of making major rectifications—which took no more than the flick of a sponge on a blackboard— whereas the engineers called for definitive choices materialized in the blueprint and in the Clinton plant: "I had to say . . . you have to stop fooling around and concentrate on one technique."[73] Despite these problems, the pilot plant somehow began to operate in late 1943, just when, on the other side of the continent, a building site of a vastly larger scope was coming into being.

The building of the pilot plant had been conceived of as a dress rehearsal for the construction of the production plant and the technical processes it

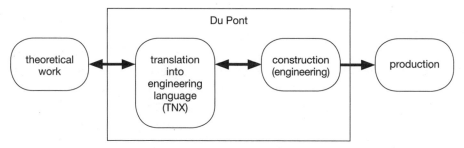

Fig. 4.2. DuPont's strategic position in the Manhattan project

would house. Which cooling system should be adopted? What kind of chemical separation? What protections against radiation? This exercise brought out different cultures collaborating in an uneasy and conflict-laden manner. Unlike in the development of nylon, where the scientists were clearly subordinate to the chemical engineers, the contours of the Manhattan Project were rather blurred: both work and power were distributed ad hoc and amid intermittent conflict, because the physicists did not intend to become integrated into DuPont's organizational arrangement. Their "honor" was at stake, Allison told Roger Williams.[74] However, it was not simply a matter of thwarted pride; at a deeper level, it had to do with control of the project or, to put it differently, the relationship between scientific and/or technical knowledge and decision-making power. The disagreement must therefore be understood here in the strongest sense of the term, that is to say, as Jean-François Lyotard has put it, "as a case of conflict between two parties that could not be fairly resolved because there is no rule that does justice to both arguments."[75] Here the outcome was that the engineers imposed their methods and their organization, in short, their own technical and organizational language, which they had perfected in the development of nylon. Nylon had given rise to a technical culture that was strong enough both to deal with plutonium and to marginalize the physicists.

Hanford

The Factory in the Desert

While the Department of Engineering was building the pilot plant, TNX was already preparing for the construction of the production plant. On 14

December 1942, a meeting at DuPont headquarters on this subject brought together representatives of the Corps of Engineers and the University of Chicago, along with Greenewalt, Ackart, and Gary of DuPont's Department of Engineering.[76] They agreed on the need for a sufficiently large and sufficiently isolated site, where large quantities of water and electricity were available, and that could accommodate six atomic reactors and three chemical separation plants (with one mile between each of the reactors and four miles between each separation plant).[77] Two engineers of the Department of Engineering, A. Hall and Gilbert Church, went to examine a few possible sites, one of them on the border between California and Nevada, near the Hoover Dam, another in central Washington State, a third in northern California, and a fourth in southern Washington State, near the small town of Hanford. This last site was chosen rather quickly after Hall, Church, and some military men went to see it in mid January 1943. Its size was 670 square miles, and it was located on the banks of the Columbia River, with its abundant flow of water, in a desertlike region at a respectable distance big cities such as Seattle, Portland, Tacoma, and Spokane.[78]

The year 1943 was essentially devoted to preparing the site and designing the plant. Three major areas were marked off, each divided into sections measuring one square mile. The first zone was to receive the atomic reactors; the second the chemical separation plants; and the third the shops for the preparation of the raw materials, uranium ore and graphite.

It was also necessary to construct some 350 miles of roads, 125 miles of railroad track (to connect with the Seattle–Saint Paul line), administrative buildings, housing, a water-drainage system, and railroad, automobile, and electrical maintenance facilities, as well as to install electric and telephone lines. Several villages in the surrounding area (such as Hanford and Richland) were expropriated and emptied of their inhabitants, whose houses were converted into offices. Several thousand prefabricated houses were put up at a sufficient distance from the plants.[79] Geological studies were also carried out in order to determine the siting of the atomic plants, which would be quite heavy because of their containment walls. All this was the job of the Engineering Department and the regional firms with which DuPont had contracted. New problems arose all the time: lack of subcontractors, shortage of labor, a very high turnover, and difficulties in obtaining supplies.[80] It seems that both the Army's and DuPont's engineers had underestimated the difficulties posed by the site when they chose it. As Greenewalt recalled:

this was done in a desert, there was no food, no facilities, and no housing. In the construction people we had everything from the most utter riffraff—anybody who could hold a saw—to people who were really dedicated to doing something for the war effort. We had thirty-nine thousand people at Hanford at the peak of the construction. They all had to be fed. An outfit called Olympic Commissary was to do the feeding. All the food came by train from the west coast. They weren't doing a very good job. People started throwing dishes and they told our people that if something wasn't settled about this they were going to demolish the cafeteria. We fired Olympic Commissary, did the thing ourselves, and had no more trouble. There were very great accomplishments.[81]

The violent wind that often blew raised clouds of dust that made life difficult for the workers, who were housed in tents for several months. Workers had to be recruited as far away as Alaska and the Mexican border. In late 1943, 25,000 people were working at the site; a year later there were 45,000.

Parallel to the building, TNX was working on the design of the future facilities. First of all it was decided, for reasons of convenience, to build reactors cooled by water rather than by helium, after long debates about the respective merits of the two processes. But this decision gave rise to concerns about corrosion, which might smother the chain reaction. This was particularly vexing, since no one knew the effects of radioactivity on corrosion phenomena. According to the calculations, 120 tons of water per minute ideally had to pass through the reactor to cool it down, but given the absorption of neutrons, only 2,000 liters were possible.[82] The water would have to be allowed to flow freely around the uranium rods, which were encased in an alloy of aluminum and magnesium (two corrosion-resistant metals). Of concern as well were possible mineral deposits in the pipes, and so a plant to demineralize the water of the Columbia River was built ("a monument that cost five million dollars," Greenewalt called it). This plant was never used, because a quicker way to purify water was found before it was finished. But all options were explored at the same time; cost was of no account. The imperative was to produce plutonium as soon as possible.

The construction of the reactors themselves began in late 1943, before the blueprints for them had been finalized. Huge buildings, veritable concrete capsules, were erected and filled with masses of graphite pierced by thousands of holes, through which the uranium rods would be inserted. All in all, there was nothing very unusual about the construction work, except that the tolerances were extremely low; the graphite bricks had to be fitted to the tenth of a millimeter. The materials were checked and rechecked before they were assembled. Inside the buildings, the areas surrounding the reactors would become inaccessible after the reactors were put into operation. A central remote-controlled crane would carry out the principal operations, such as the loading and unloading of the uranium and maintenance work, but correcting major errors of construction would then be impossible. Moreover, the buildings had to be absolutely air-tight to prevent any radioactive contamination of the atmosphere.

The chemical separation units presented fewer problems to a chemical firm like DuPont, but in 1943, only the shells of these buildings were completed; there was no rush as long as the reactors were not supplying their irradiated uranium. The separation plants were completed in December 1944, just in time to receive the first lots of irradiated uranium. Until the spring of 1944, most of the labor force concentrated on the nuclear reactors.

Loading the first reactor (reactor B) with uranium began on 18 September 1944. But before long, to the stupefaction of the engineers and scientists present, the chain reaction that was started on 26 September stopped. This meant that something was smothering the reaction by absorbing the neutrons.[83] After the hypothesis of a water leak had been briefly entertained and then discarded, calculations rapidly led to the conclusion that the "poisoning" of the reactor was caused by an isotope of xenon (^{135}Xe), a by-product released by the nuclear fission. This xenon "poisoning" has become a classic case in nuclear engineering and is well known to the experts.[84] This phenomenon had not been detected at Oak Ridge, because the test reactor there was not powerful enough (1 megawatt compared to 250 at Hanford) to cause significant quantities of xenon to accumulate. The problem was solved by inserting extra rods of uranium into the graphite. The Chicago physicists had earlier criticized the DuPont engineers' decision to make the reactor larger than needed, but it was precisely this allegedly "conservative" design—in which the systematic addition of safety margins slowed down the construction process—that in the end allowed the reactor to function properly. At the time, the engineers had not foreseen the appearance of xenon, but they were concerned about the fact that the production reactor was 200 times more powerful than that of the pilot plant and felt that this difference in scale was much too great. "We would never do it in our own manufacturing operations," Greenewalt explained. "Therefore I said, 'we're going to put in every margin of safety that we can think about—any way of preserving a margin—so that if something happens we will be prepared for it.'"[85]

This story is told in detail by Richard Hewlett and Oscar Anderson in volume 1 of the official *History of the United States Atomic Energy Commission*, and it has been repeated by all the historians of the Manhattan Project, who consider it proof that the scientists could not have successfully completed the project on their own. "More from luck than foresight du Pont was prepared to meet the situation," Hewlett and Anderson write,[86] whereas Greenewalt invokes the principle of the safety margin, even while acknowledging that he had not foreseen the formation of xenon. One might be tempted to see all of this as an obscure technical question: if the reactor had not been built larger than needed, it would have been rebuilt; the problem could have been remedied, given that the level of radioactivity was still fairly low; and in any case, a plutonium bomb would have been ready before the end of the war, pace David Hounshell.[87] But the fact is that this incident gave the engineers the opportunity to stress the essential importance of their input and to assert their preeminence. Greenewalt's diary insists that the scientists did not know what to do, that their calculations had not given them any warning, and that the DuPont men saved the day. He presents it as the triumph of the "plumbers"—the engineers—over the physicists ("with their lily-white hands," as he puts it). However, by September 1944, the DuPont engineers had been in charge of the project for some time, and so we must see all of this as a purely symbolic retribution rather than as a bid to change the existing power relations.

This story has become legendary at DuPont, to the point where someone even composed a ballad to honor "Old Marse George" (no doubt George Graves), the DuPont engineer responsible for making the reactor larger than necessary. (The "longhairs" are the physicists.)

Old Marse George
The tale's been told, as well you know,
That Hanford nearly flopped, although
The piles were later made to go
Through brilliant engineering.

The reason they were made to run
Was that a battle had been won
Long months before in Wilmington,
With brains and persevering.

We'd cobbled up a tight design
Hewed strictly to the longhairs' line.

To us, it looked almighty fine,
A honey, we'd insist.

But Old Marse George, with baleful glare,
And with a roar that shook the air,
Cried "dammit, give it stuff to spare
The longhairs may have missed!"

.   .   .   .

And later when the crisis came
Twas George's trick that saved the game,
And thrice we bowed, and praised his name
From dusk to dawn that day.

With Nagasaki smoking hot,
The news was full of tommyrot;
And generals and colonels got
A lot of decorations.

Of Old Marse George you never heard
A single solitary word,
And quietly we gave the bird
To all the celebrations.[88]

"If you'd only had a little more faith in us at the beginning," Greenewalt sighed, "my life would have been a lot more pleasant. . . . Physicists are a funny breed; there is just no doubt about it. I think they still think that this was their show."[89]

In the course of the last weeks of 1944, the two other reactors (D and F) were loaded with uranium, and in early 1945, they began to produce enriched uranium, which was then transferred to the chemical separation plants. The aluminum coverings protecting the uranium rods were removed and the rods were treated with different kinds of chemical solutions, so that by February 1945, significant quantities of plutonium nitrate could be shipped to Los Alamos. From February to August, the plants at Hanford were operating at full capacity, without any special problems. General Groves insisted that the nuclear reactors operate at the outer limit of the safety norms (for temperatures and radioactivity) to make sure that the bomb would be ready in time. Germany was defeated, and the Red Flag was waving over the ruins of the Berlin chancellery, but there was still time to try it out on Japan.

Nagasaki

By early 1945, the problem of how to produce fissile materials—plutonium and ^{235}uranium, which arrived in quantity from the isotopic separation plants at Oak Ridge—had been solved. The task at hand now was to design bombs, that is to say, to find the best way to trigger the instantaneous chain reaction of several kilograms of uranium or plutonium, which in less than a hundred millionth of a second would release an amount of energy equivalent to several thousand tons of TNT. This last stage no longer concerned DuPont but was in the hands of J. Robert Oppenheimer's team at Los Alamos.

By this time, the physicists had realized that only plutonium could be produced in large quantities, thanks to the Hanford facilities. And indeed, ^{235}uranium, produced by isotopic separation, required too large a quantity of uranium ore, and its fissile qualities were not as good as those of plutonium, whose critical mass was also smaller. "Little Boy," the bomb dropped on Hiroshima, was the only device of this type.[90] All the other atomic bombs built by the United States and the USSR after the war were plutonium bombs. In early 1945, some problems concerning the design of the plutonium bomb still remained to be solved: the physicists had chosen to trigger the chain reaction by implosion, which although more complicated than the cannon principle used in the uranium bomb worked better in practice.[91] The plutonium was placed in the center of the bomb and surrounded by sixty-four charges of conventional explosives, which produced excessive pressure on the plutonium when triggered. Suddenly compressed in this manner, the plutonium instantly reached critical mass. A polonium-beryllium detonator produced the first neutrons needed for the chain reaction. The architecture of this bomb was different from that of "Little Boy," whose interior cannon propels the explosive matter at a relatively slow speed. But then it was found that plutonium, unlike uranium, contains a small quantity of another fissionable isotope, ^{240}plutonium which can bring about a mini-explosion before the two critical submasses of ^{239}plutonium have made contact. Sparking the process by means of peripheral explosives allows for a much more rapid compression: just one-millionth of a second.

Plutonium turned out to be crucial in the context of the postwar period and the large-scale production of atomic bombs. After the war, the arms race and the accumulation of nuclear devices on both sides of the Iron Curtain had their origin at Hanford—even on the Soviet side, because their first reactor was an exact copy of the Hanford reactors, and their first bomb was the twin sister of "Fat Man," the American plutonium bomb.[92]

The plutonium bomb was successfully tested for the first time on 16 July 1945 at Alamogordo, New Mexico. On 9 August 1945, three days after the Hiroshima explosion, a B-29 bomber called "Bock's Car" took off from an airfield on Tinian island in the western Pacific with "Fat Man" in its bay. Two weather observation planes preceded it to the two potential targets, Kokura and Nagasaki. They reported good visibility over both cities, but when "Bock's Car" arrived over Kokura, the crew found a thick cover of fog, which hindered the dropping of the bomb. The B-29 therefore changed course at the last minute and made for Nagasaki. "Fat Man" exploded over that city at 11:02 a.m. At the time, Nagasaki had 250,000 inhabitants. The explosion, of a power of 21 kilotons—or the equivalent of 21,000 tons of TNT—instantly killed 35,000 people and injured 60,000 more, many of whom subsequently died.[93] The death toll was lower than at Hiroshima because of Nagasaki's steep hills.[94]

The Engineers in Their Glory

The Manhattan Project involved a process of negotiation between different idioms and different technical protocols. But the negotiation was lopsided, in the sense that after 1942, the military had chosen to grant a preeminent position to the engineers. There was nothing "self-evident" or "natural" about this—no historical process is ever self-evident—for, after all, the physicists and the Army's engineers could, with the help of Stone & Webster, have handled the industrial aspects of the project. In that case, the plutonium bomb might not have been ready by July 1945, and Nagasaki might have been spared. But not all the physicists were lacking in industrial experience. Some of them had built cyclotrons in the 1930s, as Greenewalt recognizes at one point.[95] On the Soviet side, the physicists directed operations and successfully manufactured industrial quantities of plutonium under the vigorous leadership of Igor Kurchatov and Yuli Khariton.[96] They had help from engineers, but the latter were entirely subordinate to the physicists. The chemical industry had not been a high official priority in the USSR before the war, and important areas, such as the purification of rare chemical elements and improved instrumentation, had been neglected.[97] The USSR only produced simple chemicals in large quantities—rather like the United States before World War I—and chemical engineering was less sophisticated there than on the other side of the Atlantic. The

Soviet engineers lacked the conceptual tools that would have allowed them to claim a say in the matter.

DuPont's participation should therefore not be considered a foregone conclusion, as if no other option had been possible: if this firm was chosen, it was because its engineers also exhibited a certain culture and a certain political attitude that the military engineers found to their liking. The latter were engineers who shared with the DuPont engineers a way of looking at problems, a culture of the blueprint and the technical drawing. By contrast, they were not as comfortable with the mathematical culture of the physicists.

In addition, transferring the control of the project from the physicists to the engineers provided a guarantee against any "rebellion" on the part of the physicists and limited its possible consequences. Most of the scientists had a rather suspicious, even hostile, attitude toward the military, and their participation in the Manhattan Project was primarily motivated by their anti-fascist engagement. Groves was quite suspicious of the physicists and felt that it would be safer to call on DuPont to take charge of the project. "Groves had total confidence in Du Pont—and in himself," was Greenewalt's conclusion.[98] This confidence was not only technical but political. The politically conservative DuPont engineers not only had a political culture that very closely matched that of the military, they were also socially close to them. By contrast, many of the physicists were culturally foreign to them, either because they were of European origin—especially such men as Wigner, Szilard, and initially Fermi, who were deeply suspicious of DuPont—or because their political convictions were different. I do not mean to introduce here a determinism that would deduce inflexible norms of behavior from national and social origins; nor do I mean to reify people's political identities. All I want to do is to find sociopolitical indicators that converge with the formation of the groups themselves.

The Manhattan Project gave the chemical engineers the opportunity to demonstrate their versatility, for in a short span of time, they went from nylon and the mastery of polymers and high-pressure chemistry to nuclear engineering. The development of nylon had been a determining factor in bringing together the group of chemical engineers working for the company; the production of plutonium gave this group visibility, not only within DuPont itself but also in the larger scientific community. At the end of the war, Greenewalt recalls with pleasure, Fermi proposed that he leave DuPont in order to work with Fermi in his brand-new Institute of Nuclear Physics at the University of

Chicago. Greenewalt declined on the grounds that he did not have the necessary mathematical competence, but also, and above all, because he could see dazzling prospects for his advancement at DuPont, and because his professional status could be validated most effectively within the company: "I had a career here."[99]

However, not all the engineers made this choice. Some of them used the technical baggage they had accumulated during the war to create the first courses of nuclear engineering, often in university departments of chemical engineering. James Maloney, for instance, a young engineer born in 1915 in Missouri who had a doctorate in chemical engineering from Penn State, had been recruited by Chilton for chemical research in 1941 and assigned to work with the physicists of the University of Chicago two years later. In 1945, he accepted a professorship at the University of Kansas, where he soon, at the age of thirty, became chairman of the Department of Chemical Engineering. At this particular time, five other of DuPont's chemical engineers followed his example.[100] Chilton himself left DuPont in 1960 to accept a chair at Berkeley, but later returned to the East Coast, to the University of Delaware. MIT also reacted quickly, setting up a division of nuclear engineering within its Department of Chemical Engineering in 1946. Its first-full time professor was Manson Benedict, who during the war had worked on isotopic separation (specifically through gaseous diffusion, his special area of expertise) at Kellex, the nuclear division of the engineering firm M. W. Kellogg.

But DuPont was particularly attached to its chemical engineers of the Manhattan Project. Basking in the light of their two accomplishments, they projected a flattering image to the outside world, that of simultaneously being both specialists in better living and experts in national defense.

Everyone congratulated them: Walter Carpenter, the company's president, General Groves, and the Army. They collectively received the AIChE's annual prize in 1945. Five years later, Dwight D. Eisenhower, then president of Columbia University, bestowed Columbia's Medal of Excellence on Chilton. In 1947, MGM released *The Beginning or the End,* the first film about the atomic bomb. Hollywood was interested in the Manhattan Project, and MGM and Paramount had competed keenly to make it. The film's premiere on 5 March 1947 reunited the principal participants in the project, some of whom, such as Groves, had served as consultants for the film.[101] Many engineers had been invited, among them Williams and Greenewalt, who watched actors playing themselves on-screen.[102]

Few of the chemical engineers voiced any dissent or showed any anxiety about the Manhattan Project, but James Vail, the president of the AIChE, wrote a thoughtful editorial for the February 1947 issue of *Chemical Engineering Progress:*

> It is conventional to assume that the individual chemical engineer has little influence and no responsibility for the end-use of his efforts. We are very critical of the Germans who try to detach themselves from responsibility for the terrible things done by the Nazi government. Orders came from above and that was regarded as sufficient. I do not think so, for no man can detach himself entirely from his community, or escape some measure of blame for its mistakes. I, as a member of the chemical industry, was involved in the creation of the atom bomb. I cannot escape my share in the guilt for the unprecedented outrages of Hiroshima and Nagasaki. . . . How can I escape the predicament?[103]

But such questioning remained unusual in the immediate aftermath of the war. Most of the engineers were proud of their work, and the professional reviews were filled with articles extolling their achievements. *Chemical Engineering Progress,* the official publication of the AIChE, had been founded in 1947 for the express purpose of enlarging the thematic scope of the review *Transactions in Chemical Engineering,* which published only technical articles. The new review was more ambitious and published articles on such topics as the nature of the profession and the international situation. Many stressed the versatility of the chemical engineer, who was capable of responding to a variety of needs, from those of the housewife to those of the military. And if chemical engineers had made the atomic bomb, it was pointed out, they could also invent something to protect America from an atomic attack.[104] *Chemical Engineering Progress* was in the hands of DuPont men: the president of the AIChE was Charles Stine, who succeeded James Vail in December 1946, and three years later, Stine in turn yielded the position to Thomas Chilton, who had been a member of the editorial board since 1947.

Several articles made the point that "credit for the large-scale development of atomic energy has mostly gone to the physicists." Yet "in informed circles it is well known that the theory was known before the project began and that the chemical engineers are largely responsible for its success," one asserted. And here the author added, with some exaggeration, that "the physicists who took part in this could only do so because they learned about engineering."[105]

A 1946 survey carried out by the Engineers' Joint Council records that of the total number of engineers, 9.8 percent were chemical engineers, 23.6 percent, civil engineers, 25.2 percent, electrical engineers, 23.7 percent, mechanical engineers, and 7.6 percent, mining and metallurgical engineers. This was also the first survey to compare the levels of remuneration for engineers nationwide. It shows that chemical engineers were the most highly paid. Experienced chemical engineers (i.e., with at least twenty-five years' experience) earned an average monthly salary of $825, compared to $693 for mining and metallurgical engineers, $650 for "other specialties," $604 for electrical, $587 for mechanical, and $513 for civil engineers. This also applied to beginners; chemical engineers were hired at an average monthly salary of $256 (recall that Greenewalt started at $120 in 1922), compared to $243 for civil engineers, $236 for mining and metallurgical engineers, $237 for the electrical, and $225 for mechanical engineers.[106] The survey had listed chemical engineers as a separate category since 1940, when 13,000 were counted; by 1950, there were 34,000.

DuPont's directors were not slow to promote their engineers from the Manhattan Project. Walter Carpenter, DuPont's president, was fifty-seven in 1945, and a successor for him had to be found in the remaining three years of his presidency. During the war, he had regularly met with Greenewalt, whose qualities he appreciated.[107] Greenewalt passed through the stages leading to the presidency at an accelerated pace: he was first appointed assistant director of development, then assistant to the general manager of the Pigments Department, then director of DuPont's subsidiary Canadian Industries Limited, and sent to London for discussions with the president of ICI—all of this in less than a year's time. This was not so much a matter of evaluating Greenewalt as of familiarizing him with the company as a whole and expanding his professional horizons, which until than had been strictly American. In 1946, he, along with Williams, was made a member of the Executive Committee and a vice president of the firm. The last stage was reached in January 1948, when Carpenter announced that he was stepping down from the presidency and that Greenewalt was replacing him.

Why is it that until now so little has been said about the role played by the chemical engineers of DuPont? One can identify two major reasons. To begin with, there was the considerable political weight of the physicists, which made itself widely felt after the war, when many of them, basking in the halo of their success, occupied positions of power in the federal bureaucracies and the universities. Historians have adopted the perspective of the physicists, as if matters of production had been secondary and left to bit players without any influence at all. Then, too, after the war, DuPont adopted a low profile concerning what was seen as a morally doubtful industrial adventure, for it was not eager to have ethical and political controversies interfere with its fabulous sales of nylon and other consumer goods. Only in connection with international crises—the Korean War, the Cuban Missile Crisis—did the company's directors make an effort to remind the public of the role DuPont had played in the defense of the country.

And yet, however interesting it is historically to exhume these engineers from their relative oblivion, more is involved here than doing justice to some anonymous historical figures. That would please only themselves and their descendants. It has to be done above all in order to reintegrate the Manhattan Project into the history of industry, as one of its variants. That history, of course, is not limited to DuPont, even though that company certainly was its main protagonist. We know little about the role of Kodak or Union Carbide, to name only these two, in the Manhattan Project, and it is possible that future historical investigations in this area will reconfigure our knowledge about the making of atomic bombs.

However that may be, one can reasonably say that the building of the atomic bomb was the fruit of an encounter between two historical developments that had hitherto ignored one another: on the one hand, a half-century of research in nuclear physics, which culminated in 1942 when Fermi and his team produced the first chain reaction in history in a squash court under the stands of the University of Chicago's athletics stadium; on the other, a half-century of mass production by the chemical industry, which had begun in the early years of the twentieth century with the synthesis of ammonia and the techniques of catalytic high-pressure chemistry and reached its apogee with the production of nylon in the 1930s. The fact is that the creation of the American nuclear arsenal required the work of both cutting-edge scientists and specialists in industrial production, just as nylon had done a few years earlier. In both cases, there was an encounter between scientists and engineers, which went smoothly in the case of nylon, because the chemists worked for DuPont, where everyone had a fairly well defined assignment, and where chemists and chemical engineers shared a similar scientific culture. It was not as easy when it came to plutonium, where there were profound cultural differences and institutional oppositions. But what the two encounters had in common was that the nylon and the plutonium had to be produced on a large scale, whether for commercial profitability or military effectiveness. In this sense, they constituted a persuasive paradigm in the postwar period, a paradigm that was all the more enduring because the same engineers had developed both products.

In this history, it is also important to consider the political dimension of DuPont's participation in the nuclear program, for it subsequently had a lasting impact on how Americans began to conceive of the effectiveness of political action. The key to success was an alliance among the federal bureaucracy, the scientific experts, and private partners who contributed their industrial know-how. World War II brought about the "marriage" between the public and the private sphere that the New Deal had never been able to achieve. The clearest difference between the two big projects, nylon and the atomic bomb, lies in the fact that the second blurred the political distinction between the public and the private spheres. They had not fused, of course—all concerned were careful to stress their attachment to free enterprise and the rules of the market—but their proper attributions were not as clearly defined as they had been before the war.

In 1952, David Lilienthal, former director of the Tennessee Valley Authority and of the Atomic Energy Commission, published an apologia for big companies entitled *Big Business*, in which he asserted that henceforth such companies would constitute one of the irrevocable aspects of modernity, which according to him was characterized by ever-growing needs that only such companies could satisfy. Yet when Lilienthal was director of the Tennessee Valley Authority between 1934 and 1945, he had become famous for his resolute opposition to private electric companies; as a trained lawyer, he had engaged in merciless litigation against some of them.[108] But now Lilienthal, like David Potter, the author of the 1954 study *People of Plenty: Economic Abundance and the American Character,* felt that the affluent society was in a position to solve social problems once and for all and to usher in a new era of social and political harmony in the United States.[109] The postwar era gave considerable room for maneuver to American engineers, who were flatteringly presented as guarantors of the nation's prosperity and security.

CHAPTER FIVE

The Heyday and Decline of Chemical Engineering

Immediately following World War II, the United States, which had emerged from the conflict stronger than ever, had no rival on the world stage. The American economy had proven its efficiency by responding to the war effort of the country, the "arsenal of democracy." Between 1940 and 1945, the gross national product of the United States had risen from $227.2 to $355.2 billion, a growth of 56 percent, and the last aftereffects of the Great Depression had been definitively erased. Unemployment, which had still affected 14.6 percent of the economically active population in 1940, had virtually disappeared by 1945.[1] Everything seemed to confirm the prediction of Henry Luce, the founder and editor-in-chief of *Time*, *Life*, and *Fortune*, who announced in 1941 that the twentieth century would be the "American century."

Americans seemed determined not to return to an isolationist policy, as they had done in the years following the Great War. Long before the explosion at Hiroshima, Undersecretary of the Navy James Forrestal declared: "We have the power now; we must resolve to keep it."[2] In the last months of World War II, the specter of a worldwide expansion of Soviet communism confirmed the American leaders in their conviction that the political and military involvement of the United States in world affairs should continue after the defeat of Germany and Japan. The remaining task was to define the forms of this mobilization and of what the historian Michael Sherry has called "the ideology of national preparedness."[3] Civilian consumer needs frustrated by four years of war had to be satisfied, but at the same time, it was necessary to maintain the bonds between the military, industry, and the great research universities that had been forged during the war and were perfectly exemplified by the Manhattan Project. How could all these objectives be reconciled? None of this was very clear yet in 1945, but General Eisenhower, the future architect of the "military-industrial complex" against which he was to warn sixteen years later, already thought that postwar America would be marked not only by prosperity but also by what came to be known as the Cold War.

Relatively isolated from the horrors of the conflict and for the most part filled with confidence in the future, a large majority of Americans believed that their country was capable of defending itself against the communist peril—and of protecting the free world—while also pursuing its economic growth and improving the standard of living of its population. To be sure, the fear of a nuclear apocalypse was present in people's minds right after the war.[4] But the journalists, carried away by their enthusiasm, also wrote that in the future, electricity of nuclear origin would be too cheap to be measured, that automobiles would be moved by "a few atomic grams," and that sports events would never again be interrupted because of rain. Clearly, the memories of the Great Depression, which were still alive at the end of the war, as Eric Johnson, president of the national Chamber of Commerce, remarked in 1944, gradually faded in a "society shaped by prosperity, power, and the promises of the American era."[5]

In this scheme of things, large corporations were assigned a major role in the mass production of goods for the population and for the military. Johnson summarized a widespread attitude when he asserted that "Big Business is the intrinsic expression of the era of technology." Deploring the fact that not all Americans were yet convinced of its blessings, he felt that big business was necessary, because it brought them "the fruits of mass production" while also "burying the Axis powers under an avalanche of military materiel."[6] This kind of declaration of faith in big business and its strategic and economic role was frequently heard in the postwar period, even from those, among them quite a few New Dealers, who in the past had shown a certain distrust of business.

A half-century earlier, Progressives and muckraking journalists had used strong language to denounce the rising power of the large corporations, which they saw as threats to American national values. World War II marked a major turning point, in the sense that the return to full employment attenuated the calls for social reforms at a time when the constraints of war production created a new climate of cooperation among corporations, labor unions, and the federal government. In 1945, criticism had not totally ceased, of course, but America seemed to be reconciled to its large corporations.[7] Better yet, people assigned them a central role as guarantors of prosperity and security in partnership with the public authorities.

DuPont, even more than others, benefited from this new climate. Recall the difficult position of this company before the war, when its directors, mired in a rhetoric that came straight out of the social Darwinism of the nineteenth century, had presented an odd and ambiguous image of their firm as a blend of cutting-edge technology and reactionary politics. All of this seemed to be forgotten now. The du Pont brothers had retired, and everyone expected the postwar period to be a prosperous time for the company. Sales of nylons, which had been interrupted by the war, were expected to be fabulous, and everyone said that women could not wait for them to resume. But there was one thorny question: How would the company negotiate its exit from the nuclear venture, as its directors had promised themselves they would when they had agreed to become involved in the Manhattan Project in 1942? In late 1945, DuPont was still running the Hanford facilities that produced plutonium for atomic bombs, and it became clear to its directors that the war had opened a new page: the government would more than ever be the inevitable partner of big business.

In short, DuPont was facing the political issue of the link between the company and the public sphere, and in the final analysis, this issue was not unrelated to the more general matter of the political and economic goals of the American people. This chapter examines the manner in which DuPont eventually became part of the military-industrial complex by producing the fissile materials needed for the large-scale manufacture of hydrogen bombs while remaining very largely engaged in civilian activities. These two main categories of activities were associated within the same organization and carried out by the same engineers at different stages of their careers.[8] In their close association, they presented the new and somewhat asymmetrical face of the postwar United States.

Yet this is something that historians of the postwar period have not shown in a convincing manner. Some have concentrated on the genesis and the rise of the military-industrial complex and its consequences for the American economy and society. Others have focused their attention on a civilian society seemingly gratified by the benefits of mass consumption. Historians usually dissociate these two aspects, and while some of them have intuitively perceived and suggested the connection between them, they have failed to demonstrate it. In such a perspective, the military-industrial complex is understood as a rigid system impermeable to other sectors of production. What I wish to show here is that at its margins, exchanges did take place, and that if a war economy could function for so long, it was because not all of its participants were cut off from production for the civilian market. This, then, was not a "closed world," to quote the title of a study by Paul Edwards, but rather an "open world," less isolated and more changeable than has often been described.[9]

No Real Exit from the Nuclear Venture

Nylon Comes First!

Rarely has an entirely new product, such as nylon, so perfectly matched the expectations of its consumers, especially female consumers. Rarely has a product fascinated the press to the same extent and provided it with so much material for reporting and epic stories. When the sale of nylon stockings resumed in 1946 after a five-year hiatus during which nylon was used exclusively for the production of parachutes and a few other military items, such as mosquito netting—the papers reported riots in front of stores. "Peace is finally here: nylons are for sale" exclaimed the *Chicago Sun,* no doubt by contrast with the preceding months, which were described as follows by the *Grand Rapids Herald:* "We are a free people again. But where are our nylons? That's the big question today. . . . There ought to be a law."[10] In Tulsa, a journalist asked sixty young women what they had missed most during the war; twenty replied, "men," and forty, "nylons."[11] "The shortage of nylons might drive young women into the arms of the Communists," one man in Los Angeles worried.[12] In France, women still talk about how they cut up the parachutes of American soldiers to make the blouses they wore with pride and pleasure during the penurious days after World War II.

Nylon is one of the great symbols of the American century, on a par no doubt with Coca-Cola in the consumer dreams of twentieth-century men and women. Unlike Coca-Cola, however, it is not only a technologically advanced product, it has also captured the public's imagination in the form of women's stockings and lingerie: nylon as an object of desire. When the world emerged from the somber years, it embodied the return to entertainment, to dances, to going to the movies and sauntering on the boulevards. It was the promise of a gentler life.

DuPont's essential concern was therefore to respond as quickly as possible to the civilian demand for nylon, if only to give the lie to spiteful commentators who suspected the chemical firm of purposely prolonging the shortage in order to heighten the excitement. War production was reconverted to supply the manufacturers of stockings and other textile goods. Until about 1952, most of the nylon (80%) was made into stockings and lingerie, but then other sectors of the clothing industry, such as men's underwear and shirts, also succumbed to the nylon craze. In the beginning, when production was insufficient to meet the demand, nylon stockings were available in limited quantities only. In San Francisco, for instance, the Weinstein department store had the exclusive right to sell three pairs to each customer, three and no more! Contrary to the fears of certain economists, who remembered the recession following World War I, 1946 was marked by the very vigorous growth of a civilian demand that had been frustrated during the war years. This was even more pronounced in the case of nylon, for its budding legend, which had sprung up before the war and was then passed on by word of mouth during the years of restriction, raised expectations to a particularly high pitch. Rumor even had it that in certain cities the mobsters were thinking about the possibilities of trafficking in nylons, in hopes perhaps of recapturing the illicit profits of Prohibition.

These anecdotes must of course be taken with a grain of salt. They tend to present women as naive and somewhat childish consumers, incapable of controlling their impulse buying. This image is part of a genre, and we should not be taken in by it. When it comes to nylon, true and not so true stories are inextricably mixed, for this topic was at the top of the news. Nylon was the smiling symbol of the postwar era. And indeed, military production was quickly reconverted, and the shortages were only temporary and local. As early as 1944, Henry du Pont, the company's vice president, perceptively told his employees: "Instead of being faced with the problem of spending large sums of money to convert our existing plants back to peacetime production, we are going to be faced with the problem of expanding the manufacturing capability of many of our plants in order to take care of increased civilian demand, and to maintain our share of the business."[13]

However, a few months after the explosion of the two atomic bombs in Japan, the directors of the chemical firm faced another, equally important but more arduous issue: namely, whether they wished to maintain relations with the government, particularly in the framework of its nuclear program. In principle, the matter was relatively clear, since DuPont had signed on to the Manhattan Project only on condition that it would be allowed to leave it as soon as the war was over. But the situation turned out to be more complex because of the pressure of the military and the politicians, who wanted DuPont to stay in the nuclear field.

In a first phase, General Groves asked DuPont to extend the contract it had signed in 1942 to 31 October 1946, so that the Army would have time to find a replacement. The head of the Manhattan Project saw this as the first tactical step toward persuading DuPont to stick with its nuclear activities in cooperation with the public authorities.

The directors of the chemical company had already envisaged this possibility. As early as 1944, the Executive Committee had asked itself whether nuclear work should be pursued after the war. After all, this sector held many promises. This is what Greenewalt summarized in his wartime diary, where he detailed the advantages that would accrue to the company from its presence in the nuclear field,[14] which included the prestige attached to this new technological frontier and the opportunity to exploit radioisotopes in numerous areas of industrial chemistry. Still dismayed by the trouble he had had with the Chicago physicists, however, Greenewalt doubted whether it would be possible for the company to hire these "odd birds." After all, only John Wheeler had agreed to work in Wilmington during the war (and in fact continued to do consulting work for DuPont), and except for a few other physicists—Greenewalt thought Fermi might be one of them—there was little hope of attracting them.[15]

Greenewalt also indicates that he had broached this subject with Fermi, Compton, and other Chicago scientists as early as mid 1943, but he does not explicitly relate what they said to him.[16] We know that Fermi asked Greenewalt to stay with him at the University of Chicago, where he was planning to create a nuclear physics institute, as Greenewalt often proudly recalled. But this does not mean that Fermi was in favor of having the DuPont *company* continue in the nuclear field. The physicists had come to recognize the contribution that DuPont's organizational and technical expertise had made to the Manhattan Project, but they still felt somewhat ambiguous about it. It would seem that Fermi and his colleagues hoped to secure the services of these chemical engineers independently of DuPont. The idea, then, was to bring the engineers' expertise into an academic and semi-public setting, for the projected institute was to work in collaboration with the future Atomic Energy Commission (AEC). In other words, they wanted the advantages of DuPont without DuPont.

Beyond that, the commercial prospects of a nuclear industry were highly uncertain, and huge investments would have to be made—to the detriment of other areas—before any profits could be expected. No one knew exactly whether it would be possible to exploit nuclear energy in a cost-effective manner, given the fixed investment costs. And very few experts in the nuclear field believed in the marvelous promises touted in the newspapers about electricity one day being "too cheap to be metered."[17]

On 12 February 1945, the Executive Committee, of which he was a member, had asked Charles Stine to prepare an official report on this matter. In this document, he essentially repeated Greenewalt's arguments and then added a few considerations of his own. Stine stressed that it would be unfortunate to give up the obvious advantages that would accrue from maintaining and using the technical and scientific capital the company had accumulated in a few years through the Manhattan Project, for these would not only enable it to seize commercial opportunities when they arose but also keep it in touch with the physics community. Nonetheless, he also pointed out that DuPont would have to venture into a new area, that of energy production, the only major outlet in the civilian market. But all of this would require considerable investments, the mobilization of very large numbers of employees, and above all the conduct of research, "under the punctilious supervision of the government," which might entail "possible political repercussions."[18] Here the elderly director, an associate of the du Pont brothers, seemed to show his habitual distrust of the public powers, continuing in the direct line of what he had said before the war.

One last factor, this one of a technical nature, may have been involved. At Hanford, clouds were gathering on the horizon. The physicist Eugene Wigner had calculated that the graphite of the reactor was liable to expand over time (this phenomenon is now known as the "Wigner effect"), and in mid 1946, no one knew how to prevent this troublesome development.[19] Just recently, one of the three reactors at Hanford had had to be turned off for this reason. The expanded graphite had deformed the uranium rods and the pipes carrying the cooling water. The techniques of nuclear engineering had not yet stabilized, and heavy investments were still necessary.

The Executive Committee made its decision in the spring of 1945: confirming its departure from the nuclear field, DuPont proceeded to make massive investments in new production units for nylon, paints, and various other products. The company's directors, following the recommendations of Stine, felt that the civilian market should have priority and that venturing into the nuclear field would be too risky, both economically and politically.

Despite a last attempt by Secretary of Defense Robert Patterson and by President Harry Truman, who summoned Carpenter to the White House on 15 March 1946, DuPont stuck to its decision. As Brian Balogh has shown, in 1945, no company was enthusiastic about the possibility of investing in the nuclear field, and General Electric finally took charge of the Hanford facilities in late 1946 only after long negotiations with the AEC and considerable hesitation.[20] Groves had approached General Electric as early as spring 1946, holding out dazzling possibilities of massive electricity production. But the electrical company's directors were extremely skeptical, particularly Irvin Langmuir, the vice president for research.[21] The 3,950 Hanford employees were transferred from DuPont to General Electric, and only 135 returned to DuPont. Among these 135 persons were 67 chemical engineers and 45 civil engineers.[22] DuPont thus kept most of its Hanford chemical engineers and found positions for them in other departments. A few were assigned to a nuclear "watch." The chemical company obviously considered its engineers as assets and recovered a fair number of them. (We do not have reliable figures concerning the 3,950 employees who remained at Hanford, but many of them must have been engineers.)

DuPont's retreat raised questions about the organization of the American nuclear program, a matter that was vigorously debated in 1946. The military, under the leadership of General Groves, and with the support of certain politicians, wished to keep control over the nuclear facilities, whereas the majority of the physicists, sustained in this endeavor by certain members of Congress, such as the young and ambitious Democratic senator Brian McMahon, wished to transfer the administration of the laboratories and plants to a civilian agency.[23] There was also talk of an international control of nuclear energy. In October 1946, when Congress was debating the May-Johnson bill—a bill designed to accommodate the wishes of the military—the physicists entered the political arena by creating the Federation of Atomic Scientists. Through this group they intended to make themselves heard by the politicians and to promote the civilian uses of nuclear energy, as well as the widest possible dissemination of information about it. Scientists from Los Alamos, Oak Ridge, and Chicago took up residence in Washington, D.C. No representative from Hanford was with them; according to Margaret Stahl, this was because the Hanford scientists were "closer to the business community than those at the other sites."[24]

The debate raged in the autumn of 1946, but in the end the advocates of civilian control imposed their views by noisily touting the brilliant future of civilian atomic energy and surreptitiously stoking the enthusiasm and the imagination of the journalists. The McMahon bill, passed on 20 December 1946, eventually became the Atomic Energy Act, which in turn led to the creation of both the AEC and a congressional committee charged with overseeing atomic activities, the Joint Committee on Atomic Energy (JCAE), under the chairmanship of McMahon. The military had lost the legislative battle, but a number of issues had not been resolved, particularly concerning the functions and the priorities of the AEC. Even if DuPont had made its decision before the Atomic Energy Act was passed, the intensity of the controversy was not apt to encourage its Executive Committee to remain in the nuclear field. The "political repercussions" evoked by Stine referred to this intensely political context, in which the public's demands were not clearly formulated. Who would direct these nuclear activities? What would be the link between the military and the civilian authorities? And for what purposes? Unable to provide clear answers to these questions, not even President Truman himself could persuade Carpenter.

The Era of the Nuclear Watch

From this point on, the company's "official" story as disseminated among its shareholders, its employees, and the public—and by and large endorsed by David Hounshell and John Smith in their 1988 book *Science and Corporate Strategy*—practically no longer mentions DuPont's nuclear activities. To be sure, no one denies that the company returned to them after 1950, but this venture is evoked only briefly, as if it had been a sideline hardly worthy of notice. But the fact is that two aspects of this official discourse appear to be questionable: the genuineness of the cutoff of 1945 and the secondary character of DuPont's return to military nuclear activities. Perhaps we should compare these two discourses—that of the historian and that of the company—not in order to find a middle way between them, but because DuPont, like many other American corporations of that era, had to live with a difficult tension between its military and its civilian activities. This tension is precisely reflected in the difficulties one encounters in reconciling the official with the historical discourse. Let us begin, then, with the official story.

The idea of international control of nuclear energy evaporated rapidly in the course of the year 1947, when relations with the USSR deteriorated beyond repair. The American military and the politicians now came to feel that the stock of tactical and strategic atomic weapons must be enlarged as a means to meet any Soviet attack in Europe or in Asia. The so-called containment doctrine gave a special place to the bomb, which had the potential to reverse the power relation in favor of the "free world." At that point, the United States only had a few bombs, and even these were only partially assembled. At the same time, the newly created CIA estimated that the USSR would not have a bomb available before the beginning of the 1950s, probably by 1953 or 1954. These predictions were again conveyed to the American government on 1 July 1949, a few weeks before the explosion of the first Soviet bomb on 29 August 1949.[25] The radioactive cloud that spread that day was detected on 3 September by a specially equipped American airplane, and President Truman officially announced the news to the Americans on 23 September, once he had gotten over his initial surprise and incredulity. Truman and his secretary of defense, Louis Johnson, had trouble believing that the Soviet Union was capable of catching up with the United States so quickly.

The news prompted the members of the congressional JCAE to demand a "total effort" to develop the hydrogen bomb, even before Truman had ever heard of it. This was on 6 October 1949, five days after the People's Republic of China was proclaimed—an event perceived in the United States as one more proof of Stalin's territorial ambitions. In October 1949, the military lobby set out to push for the "super bomb" as it was called, which would be a thousand times more powerful than the atomic bomb. The physics community became deeply split over it. Some scientists and administrators of the AEC were willing to work on it (Edward Teller, Ernest Lawrence, Lewis Strauss, Gordon Dean), others were not (Hans Bethe, Enrico Fermi, James Conant, David Lilienthal), and still others could not make up their minds (J. Robert Oppenheimer, Leo Szilard).[26]

One last factor encouraged the advocates for the new bomb: Britain's disclosure of the espionage activities of the physicist Klaus Fuchs, who had been arrested on 3 February 1950. This seemed to give credibility to the idea that the USSR had the atomic weapon only because of the treason of some scientists at the Manhattan Project. On 9 February, an obscure senator from Wisconsin, Joseph McCarthy, asserted in a widely commented speech given at Wheeling, West Virginia, that he had a list of 205 Communists working for the State Department. Was it not time for the United States to recover its preeminence by going from the atomic to the "super" bomb? And yet the nuclear arsenal of the United States was already formidable in 1950: the Americans had 298 fully assembled bombs available, while the USSR had five.[27] In addition, the Strategic Air Command, headed since 1948 by General Curtis LeMay (later known for his wish to "bomb the Vietnamese back to the Stone Age" and for having run for the vice presidency on Barry Goldwater's ticket in 1964), had 868 bombers by early 1950, among them 250 B-29, B-36, and B-50 bombers, which were specially modified to permit the release of atomic bombs.[28]

According to Lilienthal, it appears that Truman made his final decision very early in 1950. The president believed that the hydrogen bomb should be used to exert pressure in international negotiations, a position that was followed until 1980 by all of his successors, who always asserted that they were opposed to the use of this weapon, even though it was produced in large quantities. Also in early 1950, the AEC began to think about the design and production of the new device.[29] As had been the case in 1942, the question arose as to who should be the industrial operator of the project. It could have been General Electric, but that company was already running the Hanford facilities, and its experience was still limited. It was thus only logical that Sumner Pike, the interim administrator of the AEC—Lilienthal had resigned—should think of DuPont instead and send a letter to this effect to Greenewalt, president of the company since 1948, on 12 June 1950.[30] A few days earlier, on 8 June, Truman had decided that the program would be rated "extremely urgent" for the country's defense.

In his letter, Pike stressed that it was important that the program be placed under the responsibility of a single organization, adding "we feel that the Du Pont Company is uniquely qualified to undertake complete responsibility for a job of this magnitude with assurance that it would be completed in the shortest practical time."[31] For the moment, he simply wanted to establish a relationship by specifying the nature of the work. This might then serve as the basis for a contract to be signed between DuPont and the AEC. The Executive Committee approved this demand on 16 June 1950 and on the same day set up an Atomic Energy Survey Committee, charging it with reporting on the facilities of the AEC.[32] The eleven men on this committee had all worked for TNX.

The resolution of the American government was strengthened still further after 25 June, when North Korean troops invaded South Korea. A month later, Truman again wrote to Greenewalt to reassert the vital importance of the project and make it clear that DuPont, given its experience with Hanford, should take it over.[33] At its meeting of 26 July, the Executive Committee discussed Truman's letter and the report of the evaluation committee, which recommended DuPont's participation in the project and formulated a few technical recommendations. On that same day the Executive Committee adopted the following resolution: "The Explosives Department is authorized to express to the AEC Du Pont's willingness to undertake a new project for the AEC, on a cost replacement plus one dollar basis."[34]

On 4 August, Greenewalt, at his own request, spoke at a hearing of the congressional Joint Committee on Atomic Energy. He started by declaring: "The Du Pont Company did not seek this assignment. It was accepted upon assurances from highest governmental sources that the project is of vital importance to the security and the defense of the United States." "Much as we deplore the international unrest that makes this project necessary," the chairman of the chemical firm added, "it is nevertheless a source of satisfaction and pride [for us]."[35] This declaration was rather more cordial than that of 1942, when Du-Pont had signed on to the Manhattan Project; it was a first—and subsequently confirmed—indication that the return to the nuclear field and the alliance with the AEC were actually not unwelcome to the company's strategists. The negotiations leading to the signing of a contract took several months, because of disagreements over social benefits for the DuPont employees assigned to the project, and also the responsibility for insurance payments: mere side issues, considering the circumstances.[36] But even before a final agreement was initialed in July 1950, DuPont's managers began their preparations for honoring the contract.

This, then, is the official story of DuPont's return to nuclear activities as found in the history of the AEC and the documents disseminated by the company. The historian must consider it with some caution and wonder about a chronology that makes DuPont appear as a passive partner, buffeted by history, and forced by a critical juncture of international circumstances to go along with the political choices of the government. A better approach is to try to identify the factors that made DuPont act on its own behalf within real, albeit narrow, margins of freedom.

In order to justify the company's active return to nuclear activities, Du-Pont's directors have always claimed that the imperatives of national defense left them no more choice than they had had when they signed up for the Manhattan Project in late 1942, and that their priorities were clearly those of the civilian market. What about these claims?

It is true that commercial activities were paramount for the company after the war. Its sales figures rose from $611 million in 1945 to $649 in 1946, 783 in 1947 (+21%), 969 in 1948 (+24%), 1,050 in 1949, 1,279 in 1950, 1,531 in 1951, and 1,602 in 1952. About 50 percent of this growth came from the growth in sales, which was made possible by the expansion of the firm's production capacity, with the other half brought about by higher prices. During the war, DuPont had spent $34.7 million on the construction of new plants or the expansion of existing ones, among them the nylon plant at Martinsville, West Virginia; this was about one-tenth of the cost of the Hanford plant. In 1947, construction costs—for a second ammonia plant at Sabine, Texas, for instance—rose to more than $116 million, and this effort continued over the next few years, with between $110 and $135 million invested annually. The number of employees also increased, rising from 67,500 in 1945 to 86,800 in 1951.[37] These considerable investments were motivated by a strong demand for consumer goods and equipment; General Motors alone bought 15 percent of DuPont's output of such products as rayon, simulated leather, lacquers, and paint.

Military activities were given short shrift. For the year 1947, which represents an average for the years 1946–50, the sales of powder for military and sporting weapons amounted to 0.3 percent of all sales (they had been 2% in 1939), those of explosives, to 5 percent (also 2% in 1939), and this included sales to coal mines, which consumed large amounts of dynamite. Meanwhile the "textile" category—essentially rayon and nylon—accounted for 23 percent (22% in 1939).

It was thus a reasonable assumption that the company had definitively returned to its habitual activities and that it continued to keep its distance from the federal government, a strategy that had been initiated early in the century, when DuPont had diversified by producing for the mass market. At this time, other industrialists echoed DuPont's attitude, for instance Frank Jewett, chairman of AT&T, who made it clear that he wished to return to the cozy prewar relationship between industry and academia. The years 1945 and 1946 were a time of questioning and readjustment for business, whose most eminent representatives hammered away at the idea of a return to normalcy, claiming that the war had been no more than a historical parenthesis.

But the fact is that a page had turned, for DuPont as for others. The federal government had imposed itself as an inescapable partner, primarily in military production, but also in large-scale science projects and economic regulation. This was something that Du Pont, like all other big companies, had to reckon with. If we see the withdrawal of 1945 as a major break in the history of Du-Pont, we shall be unable to grasp the continuity of the new relations between DuPont and the federal government initiated through the Manhattan Project, which contributed to shaping the new face of postwar America. It would be an exaggeration to claim that World War II constituted a radical turning point in this respect, for the federal government had assumed ever greater power since the end of the nineteenth century, and DuPont had had continuous relations with the public powers from the time of its founding. But its gradual disengagement was irreparably compromised during World War II and then the Cold War. From 1945 on, DuPont—along with all other major American companies—had to rethink its linkages to the public sphere.

At Du Pont, this work of adjustment was actually carried out by the chemical engineers. At the end of the war, buoyed by their military and commercial achievements, they had reached their professional maturity and could lay claim to leadership positions within the company and at the same time serve as intermediaries with the public sphere. From 1945 on, they served the double function of representing their company in dealing with public agencies and providing expertise for nuclear reactors.

The fact is that DuPont did not retreat from the nuclear field but rather was involved there in a different mode. From 1946 to 1950, it made its expertise capabilities available to the AEC, for this was a time when the lack of experts was keenly felt and when the tens of thousands of engineers and physicists who were being trained in the universities were not yet available.[38] This aid took the form of the inclusion of DuPont engineers on different advisory boards from the very beginning of the AEC. The new agency was responsible for all of the country's nuclear activities, which had hitherto been administered by the Manhattan District of the Army Corps of Engineers. (The official date of the transfer was 1 January 1947.) Three of DuPont's chemical engineers played an important role in the AEC, and many others intermittently provided their services.

The first was Hood Worthington, who as a direct associate of Greenewalt's had been in charge of the construction of the Hanford reactors; he had worked on nylon before the war: it was a classic DuPont career track. On 12 December 1946, Worthington was appointed to the General Advisory Committee (GAC) of the AEC, which included eight other persons charged with advising and supervising the commission under the chairmanship of J. Robert Oppenheimer.[39] The GAC was an important body, and the historians of the AEC have devoted a great deal of attention to it. The committee was divided into three subcommittees (Research and Development, Materials, and Production), with Worthington as a member of the Production subcommittee.[40] In this capacity, he worked in particular on the technical problems the Hanford reactor was facing just then, such as the dilation of the graphite and also the chemical separation of plutonium. The process of chemical separation had proven rather inefficient, radioactive fumes escaped from the separation units, and large quantities of highly radioactive liquids had had to be stored in huge underground reservoirs.[41] Worthington proposed the use of a different separating process, the "redox" process, which made it possible to recover not only plutonium but also uranium and other radioactive products from the irradiated rods. This process was more economical, in that it reduced the need for storing intensely radioactive by-products.

On 16 April 1947, Worthington and other members of the GAC met with representatives of General Electric and asked them to adopt these measures.[42] GE was more interested in investing in nuclear energy for civilian use and conducting research on the production of energy than in renovating the Hanford facilities, but it grudgingly agreed to do so. At the time, the GAC's priority was to increase the production of plutonium and to perfect America's nuclear arsenal. It appears that the GAC was somewhat concerned about the manner in which General Electric managed Hanford. Operations were carried out too slowly, as if the company's capacities were insufficient for the task.[43]

At that point, the AEC's activities were almost exclusively focused on the nuclear arsenal. In its generally highly technical discussions, the GAC debated the virtues of different expedients, such as adding a small quantity of tritium to the bombs in order to "dope" the atomic fission. Worthington participated actively in the meetings of the GAC, but it is impossible to determine exactly where he stood when the discussion became heated.[44] The DuPont directors followed the AEC's activities quite closely, and in a sense Worthington was the company's watchdog within the GAC, the man who enabled DuPont to retain its connection with the nuclear field, as Stine had urged in his 1945 report. Keeping a fairly low political profile, Worthington only spoke up about technical questions. He was on assignment for his company.

The second manager-engineer to act as an expert was Greenewalt himself. Although already on the launching pad toward the company's presidency, Greenewalt was appointed in 1947 to the atomic energy committee of the Joint Research and Development Board (JRDB), headed by Vannevar Bush, one of the many military institutions engaged in scientific research at the time.[45] Greenewalt was asked to prepare a report on the development of new nuclear reactors. In this connection, James Conant, president of Harvard and chairman of the JRDB's committee on atomic energy, who was somewhat alarmed about the Air Force's grandiose plans, asked Greenewalt on 10 March 1947 to look into its Nuclear Energy for the Propulsion of Aircraft (NEPA) program,[46] the goal of which was to build a nuclear-powered long-range strategic bomber able to operate in Soviet airspace for protracted periods.[47] At this juncture, the promises of nuclear energy were such that both the Air Force and the Navy, determined to be part of this sector, came up with sometimes preposterous projects.[48] Research in this area began as early as 1946 under the aegis of the military and the aircraft manufacturer Fairchild.[49] The major issue was safety. What would happen if there was an accident? Greenewalt asked the Air Force people to report on this matter, and in agreement with Conant and Oppenheimer came to the reasonable conclusion that the project was doomed to failure.[50] The Air Force did not back off right away, but NEPA definitely lost momentum. Even after Greenewalt was appointed chairman of DuPont a year later, he continued his consulting activities and was regularly asked (often by telephone) for his opinion about specific aspects of the nuclear program. Given his great interest in nuclear affairs, he was glad to act as a consultant.

The third major representative of DuPont's expertise was Donald Carpenter (no relation to Walter Carpenter), a chemical engineer who at the time was vice president of Remington (a DuPont subsidiary that manufactured weapons and munitions), and who had worked at DuPont until 1945.[51] Carpenter became chairman of the AEC's military branch, the Military Liaison Committee (MLC), on 1 April 1948, replacing General Lewis Brereton. This appointment was not a simple matter: Carpenter, although initially reticent, accepted after Lilienthal pointed out to him that his interpersonal skills might improve the sometimes difficult relations between the civilians of the AEC and the military and also appealed to the higher interests of the country. James Forrestal, now secretary of defense, actually called Greenewalt, asking him to intervene with Carpenter.[52] The latter finally accepted the appointment and moved from Wilmington to Washington. Personal decisions counted for as much here as the given structures.

As a good DuPont manager, Carpenter devoted his first weeks at the AEC to reorganizing his department in order to implement a general strategy he had worked out in cooperation with Lilienthal.[53] The idea was to strengthen the effectiveness of the MLC in order to avoid having each of the services—Army, Navy, Air Force—set up its own nuclear organization, independent of the AEC.[54] In order to accomplish this, Carpenter asked each of the three services to delegate a superior officer to the MLC with decision-making authority and organized the MLC into working sections, each of them in charge of one aspect of military nuclear activity, and each run by one military and one civilian official.[55] This reorganization into working units (to replace the division into military branches) was reminiscent of the internal reorganization of DuPont in 1921. Appointing "plenipotentiary" officers of high rank—brigadier generals or vice admirals—guaranteed that the MLC would not be a simple registering board before which some subaltern officers would appear to pass on the decisions made by the several general staffs.

Moreover, Carpenter felt that the general organization of the AEC, as developed by Lilienthal, was too centralized. All problems were passed up to Lilienthal or his general manager, Carroll Wilson, so that they were too busy to devote enough time to important decisions. Carpenter therefore wrote a report, which he handed in on 29 May 1948 and in which he strongly recommended the appointment of divisional managers, each of whom would be responsible for one well-defined function and in charge of coordinating all activities in his area.[56] Not all of Carpenter's proposals were implemented, on the grounds that decentralization might compromise security.[57] Yet in the end a limited decentralization was adopted (fig. 5.1). Thirteen divisional managers were created and the commission's bureaucracy was enlarged—its Washington staff increased from 361 in August 1947 to 699 a year later.[58] Here again, the organizational model was that of DuPont. Carpenter's diagnosis of the AEC was strangely similar to Pierre S. du Pont's assessment of DuPont in 1920, which had asked the directors to devote more time to the company's general strategy and less to its day-to-day management.

DuPont's organizational methods spread not only throughout the private business community but also within the federal government, where they were propagated by managers such as Carpenter who temporarily went from the private to the public sector. The reorganization of the MLC and then of the AEC appear to be good examples of the "modernization," as Carpenter called it, of what was in fact a brand-new public agency. To be sure, one must guard against considering this type of mutation all-pervasive. Yet it is clear that in the postwar period, many American public agencies, particularly the most recent

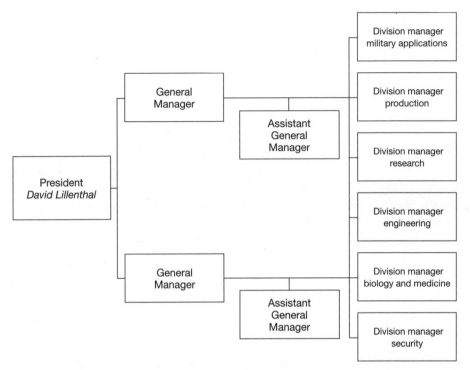

Fig. 5.1. Organizational chart of the Atomic Energy Commission after its reorganization in 1948 by Donald Carpenter based on the decentralization by function favored by DuPont managers

ones—perhaps because certain bureaucratic habits had not yet become too deeply anchored—were influenced by the management model of the large corporations, foremost among them DuPont. In this specific case, there were obstacles to overcome, for the military objected to this reorganization, although they eventually submitted to it, knowing that Carpenter had the support of Forrestal and also served as his ex-officio advisor for nuclear affairs.[59]

The best-known case of a manager entering the public sector is perhaps that of Robert McNamara, "a walking IBM computer," to cite Barry Goldwater's cruel quip. McNamara was hired by Ford in 1946, along with a group of other young managers equally keen on operations research. He made his way to the top of the Ford hierarchy and was named chairman of the company in 1960—the first who was not a member of the Ford family—before becoming secretary of defense for John F. Kennedy and Lyndon B. Johnson, in which capacity he both waged the Vietnam War and reorganized the Pentagon.[60] Carpenter's task and his profile were more modest, but in the end rather similar to McNamara's. We still do not know much about these exchanges of expertise and know-how between public and private institutions, and in particular about the very rapid dissemination of operations research as a mode of managing large-scale projects.[61]

Other DuPont experts were regularly asked to evaluate new models of nuclear reactors or to investigate matters of plutonium production.[62] The AEC's six "operational sites" (Los Alamos, Brookhaven, Oak Ridge, Hanford, etc.) directly contacted the experts they needed. For instance, on 28 October 1947, the Brookhaven National Laboratory, located on Long Island, which had just been created by the AEC for the purpose of building a test reactor, asked the DuPont engineers for an expert opinion on the use of electronic apparatus. The manager of the Engineering Department replied immediately, inviting the Brookhaven physicists to meet with him and his colleagues in Wilmington.[63]

In early 1949, as the international situation deteriorated, the AEC asked DuPont for a report on plutonium production, as well as for recommendations on this subject. At that time, General Electric was having trouble producing plutonium in sufficient quantities to satisfy the extremely high demands of the military. The Executive Committee complied, "although this does not mean that DuPont will become involved in the program that might be undertaken in consequence." Everyone remembered that DuPont had become involved in the Manhattan Project six years earlier precisely in consequence of an evaluation that it had been asked to undertake. The 1949 report stressed above all the need to find a better way to recover the radioactive elements left over after the chemical separation of plutonium, considering that these elements were stored in liquid form in underground reservoirs. These observations were very much in line with Worthington's recommendations. As a consequence of the DuPont report, written by R. Monte Evans, a veteran of Hanford and now the company's chief engineer, the GAC decided to undertake studies for the construction of new reactors, to be cooled and moderated by heavy water [deuterium oxide, or ^2H_2O] instead of ordinary water and graphite. These would be able to produce more plutonium, as well as tritium, which the physicists now considered necessary for the hydrogen bomb.

The DuPont engineers' expertise in fact suited the needs of the government between the end of World War II and the beginning of the Korean War rather well. In this transitional period, military expenditures decreased—the military budget went from $81.6 billion in 1945 to $10.9 billion in 1948—while the civilian federal institutions had not yet reached their cruising speed. If the Cold

War had not worsened, the massive administrative apparatus that regulated the relations between public and private partners might have withered away and government expenditures for military and civilian research might have returned to their prewar levels.

The Korean War brought a spectacular growth of military budgets and the birth of a permanent war economy. The fact is that a significant share of the country's economic activity—accounting for about 10 percent of Gross Industrial Product—was financed by military budgets, whose cumulative amount between 1946 and 1969 was close to $1,000 billion.[64] What now took shape, nourished by this financial manna, was what Dwight D. Eisenhower in his farewell address of 1961 called the "military-industrial complex," that is to say, a network bringing together public (the Pentagon and its various agencies) and private (corporations, university laboratories) partners. Eisenhower realized that with the Korean War, the United States had entered a new era of its history.[65]

The war and the violently anti-communist context of the 1950s could only strengthen the close relations between the public and private spheres. Consequently, if DuPont and its engineers progressed from acting as expert consultants to the more active role of principal private partner in the manufacture of the hydrogen bomb, this happened according to a timetable that precisely matched the rise in America's military expenditures and the consolidation of a permanent war economy. In this intermediary period between World War II and the Korean War, consulting relationships assumed a different character, because the traditional borders between the public and private spheres had become blurred. At DuPont, the expert advisor to the government was now a familiar figure. Because of their participation in the Manhattan Project, it was natural that the chemical engineers should occupy these strategic positions, which in turn enhanced their careers within the company. When Donald Carpenter hesitated to join the AEC, Greenewalt called him into his office and was able to persuade him. What did he tell him? Aside from appealing to his patriotic sense of duty, he pointed out that a stint in Washington would do no harm to his career—quite the contrary. Upon his return to Remington, Carpenter went back to his vice presidency, which a few years later turned into the chairmanship and a seat on DuPont's Executive Committee. His detour through the AEC had been to his advantage.

Moreover, DuPont's managers had developed a more astute political savoir faire. The AEC's committees did not function like DuPont's, and the internal relations in the commission were fraught with political pitfalls. Carpenter fell into some of these, as he did on the occasion of a visit to the White House in the company of Lilienthal. Unfamiliar with these kinds of events, Carpenter behaved as he did at meetings of DuPont's Executive Committee, where managers who were called in were expected to read reports. But, alas, no one had bothered to tell him that President Truman had little patience with technical reports, and he was brusquely interrupted. Lilienthal, an experienced habitué of the Oval Office, saved the day with his consummate savoir faire.[66] Carpenter saw this mortifying incident as Lilienthal's way of showing him who was in charge of the commission. Similarly, Carpenter's proposals for the reorganization of the AEC were not presented to the GAC in the best possible manner. But Carpenter adapted, just as Worthington had adapted to the ways of the GAC. Following Greenewalt's example, they learned how to win their cases by enlisting the support of certain partners and how to be very subtle in negotiations about a project's construction; in short, they had come to terms with the essentially political character of contemporary big science.

We know that DuPont began to prepare for the task ahead as early as July 1950. The company set up a "study group" of Hanford veterans who would work with the AEC to specify the construction and production goals that DuPont was expected to meet. The contract signed between the AEC and DuPont on 17 October 1950 provided for the construction and operation by the chemical firm of a facility consisting of five nuclear reactors, as well as several plants intended for chemical separation and the production of heavy water.[67] The unusual feature of this contract, as of that of 1942, was that DuPont would not reap any profit from carrying it out, since it would only be reimbursed for the costs incurred and paid the symbolic sum of one dollar per year. Here again, the company wished to avoid any distressing accusations of being a "merchant of nuclear death."

Why did the chemical company agree to return actively to the nuclear field when no direct financial gain could be had from it? Here we must look into a combination of factors, some of them internal and some of them external to the firm. The most important element was the political decision to increase the nuclear arsenal and to develop and produce hydrogen bombs requiring enormous quantities of fissile materials. The Korean War had just broken out and patriotic pressure was strong, as it had been in 1942. Without the express request of the government, DuPont would not have become involved in this project. The company had better things to do in a rapidly expanding civilian market. On the other hand, DuPont had kept up a "technological watch" on nuclear affairs between 1945 and 1950, thereby retaining the option of actively returning to them. It should be added that in 1950, there was none of the reticence and hesitation that had characterized DuPont's entry into the Manhattan Project in 1942. The agreement between the chemical firm and the AEC was easily reached; it was as if DuPont's directors were not displeased to be definitively engaged in the nuclear area. To be sure, the financial prospects of nuclear activities were as yet highly hypothetical, and no one in Wilmington was really concerned about them. But signing the contract did not require any soul-searching, for in a time of national emergency—and whether the United States was threatened or not is beside the point here—the country's human and technological capital had to be used to its fullest.

Within the logic of the Cold War, all of the country's technical and scientific strengths had to be used in the service of national defense, and an asset as considerable as DuPont's chemical engineers could not possibly be left out. This was something that DuPont's directors—Greenewalt in particular—recognized immediately. In his response to President Truman, Greenewalt indicated that his company would "as always," "spare no effort."[68] Like all the other great American business and academic institutions, DuPont embraced the mobilizing logic of the Cold War. After all, too, this commitment might actually pay off in the long run, some members of the Executive Committee supposed.

DuPont was then building plant after plant to satisfy the seemingly insatiable demand for nylon. Between 1945 and 1962, the production of nylon jumped from 1,500 to 600,000 tons per year. Taking advantage of this momentum, DuPont launched other synthetic fibers as well: orlon in 1949, which it presented as a "synthetic wool," dacron in 1953, and lycra in 1962, as well as terylene, trelenka, and crimplene, while its competitors brought out courtelle (Courtaulds), acrilon (Monsanto), and tricel and celcon (Celanese). All these flashy names displayed a scientific modernity that suggested the mysterious and beneficent presence of the domesticated atom and seemed to befit the atomic age. The great Parisian fashion designers were not slow to take an interest in artificial fibers either, for by the mid 1950s, Dior, Jacques Heim, Lanvin, Patou, Chanel, and Givenchy were all presenting collections of evening gowns, coats, and leisure clothing that showed every possible mode of using synthetics.[69]

Just ahead of an antitrust action by the Attorney General's Office, Greenewalt decided in 1949 to license DuPont's nylon rights to the Chemstrand Company, which began to produce nylon in late 1950, and later to foreign companies, such as the French Société Rhodiaceta, which obtained a license in 1954.[70] The fact is that DuPont was unable to meet the demand, and its directors therefore felt that the company would not lose market share by issuing licenses.[71] In addition, a major improvement program for the nylon plants was launched in 1955 in an effort to lower production costs. The plants were entirely revamped under the "ETF" program, which introduced new polymerization techniques permitting the use of continuous-flow processes.[72] Young chemical engineers were hired by the hundreds to work in the new Department of Textile Fibers, the most flourishing and dynamic in the company, on improvements in production or on new products. Their elders had created a large opening, through which ambitious young men now poured in.

The Savannah River Plant

What was the nature of the new Labor of Hercules that the AEC asked DuPont to perform? Essentially, it was to build facilities for producing heavy water, as well as nuclear reactors and chemical separation plants for plutonium production. When the negotiations for the contract were taking place, the physicists were not sure whether a hydrogen bomb could some day be built. Certain scientific problems related to fusion seemed almost insurmountable. The projected facilities would therefore have to be able to produce a varied gamut of fissile materials needed to keep up the stocks of plutonium for atomic bombs and possibly to produce hydrogen bombs.

The construction projects had a number of points in common with those that had been carried out at Hanford during the war. To begin with, an organization had to be set up. The first priority was to find a site whose characteristics were similar to those of Hanford: geologically stable, very large, far from major urban centers, and irrigated by a stream. Having studied about a hundred possibilities, the Department of Engineering opted for a site of 300 square miles located on the banks of the Savannah River in South Carolina, near the border with Georgia and slightly south of Augusta.

DuPont's Atomic Energy Division

On 1 August 1950, the Executive Committee created a new division, called the Atomic Energy Division (AED), directing it to build and manage the atomic facilities projected by the Atomic Energy Commission.[73] The head of the new division had the rank of assistant general manager and was formally placed under the general manager of the Explosives Department. He was R. Monte

Evans, who during World War II had been in charge of the manufacturing division of TNX, having earlier worked with Roger Williams in the Ammonia Department. But in practice, just as in the case of TNX, the AED had the rank of an autonomous department. Evans reported directly to the Executive Committee, without having to go through the general manager.[74]

His principal co-workers were also former Hanford men, and some had even come there from the Nylon and Ammonia Departments, while others were too young to have been hired by DuPont during the war. In addition, a head accountant and his assistant were transferred to the AED to handle the financial advances paid by the AEC. Unlike in the Manhattan Project, the government paid DuPont in advance—although adjustments to correct the initial estimates could always be made—rather than reimbursing it when the work was done. The reason was that the Savannah plant cost much more than the $350 million spent on Hanford: more than a billion dollars for the years of intensive construction, from 1951 to 1954 (successively, $24, $334, $458, and $266 million),[75] the equivalent of DuPont's sales for the year 1949. The accounting aspect of the AED was extremely complex, for the AEC, the owner of the Savannah plant, and DuPont, the contractor, each had their own accounting resources, whose methods of calculation did not match.[76]

The AED and TNX were similar in their organization, except that to handle relations with the many subcontractors working with DuPont on the Savannah project, a supervisory subdivision had been added to the AED (fig. 5.2). Even though a secret defense clearing might have to be obtained, the AED could subcontract a large part of its construction work to medium and small companies.[77] This was a rather clear-cut difference from Hanford. TNX had been a secret division, which was unknown to all who were not directly involved—and most of DuPont's employees learned about it only after Hiroshima—whereas the AED was an official division, listed by name in the company's organization charts and mentioned in publications for employees and shareholders.

The fact that the AED was an ordinary division of DuPont was not without importance, to the extent that it could call on the company's other divisions and departments—Chemical Research, Personnel, Legal Affairs, Public Relations, and Purchasing—in keeping with regular procedures. An examination of the relations between the AED and DuPont's other departments shows that a flood of services and information circulated among them, so that the AED was fully integrated into the structure of DuPont.[78] As had also been the case with TNX in the Manhattan Project, the department it most frequently called

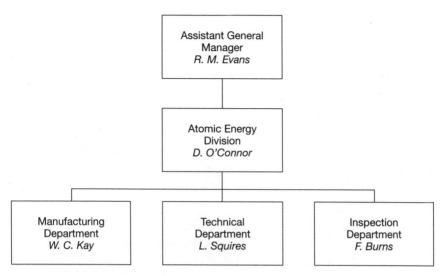

Fig. 5.2. DuPont's Atomic Energy Division, 1950
The AED was structurally similar to the TNX division of the Manhattan Project, but it was not secret. It was officially included in Du Pont's organization chart.

upon was Engineering, which was in charge of building the facilities according to the plans drawn up by the AED. This department set up an "atomic energy" working group within its Design Division. This group of some thirty engineers worked in Wilmington in liaison with another working group in the Engineering Department, the "atomic energy construction" group, which worked at the Savannah building site. The procedures for recruiting and transferring personnel between divisions and departments were by and large identical to those of the commercial divisions.

One important issue was the recruitment of personnel. The Atomic Energy Survey Committee early on asked the managers of the manufacturing and technical divisions to draw up lists of their needs for technicians and engineers. Since these managers were all veterans of Hanford, the lists they submitted were based on the experience at Hanford and on estimates. They were then examined and verified by the general manager of the division and his assistant. On 1 September 1950, the estimates stated that 900 to 1,000 engineers and technicians would be needed for 1953, 300–330 of them to work in research and development and 600–670 to supervise production. Also on 1 September, the report identified an immediate need for 112 employees in the technical division and 37 in the manufacturing division, at a time when there were only 37 employees in the entire AED.[79] Over the next few months, these estimates were adjusted, and by December 1950, the number of engineers and technicians was about 1,200 (out of a total of 5,047 employees).[80]

There was, however, some difficulty meeting this great demand for qualified personnel. Despite the Korean War, business was flourishing. The AED therefore had to compete for employees with other divisions and departments of the company. Because of DuPont's decentralized structure, the Executive Committee could not oblige the manager of a department give up some of his best employees, but there were ways of bringing pressure to bear: encouragement, reminders, and admonitions were conveyed in letters to general managers who balked at releasing a specific technician or engineer.[81] Every department of DuPont had a quota of employees to furnish to the AED and was expected to fill it to the best of its ability. All departments, without exception, were tapped.

But the company also recognized that "the job could not be done by men lacking ability and enthusiasm for the work."[82] In no case was an employee forced to work for the AED.[83] Francis Vaughan, for example, who had worked with Roger Williams in the Ammonia Department and then at TNX, was asked but refused, for he was just then working on a long-term project in the Polychemical (formerly Ammonia) Department. Vaughan felt that he could more usefully work with a new plastic, the future delrin (a commercial fiasco). The company had to make sure that careers were not harmed by a stint with the AED, that working for the government was not a dead end, and that a position in the AED carried as much prestige as a commercial one. Salary scales and benefits as well were aligned with those of the other departments. In certain cases where the AED was particularly anxious to secure the services of a specific employee, the general manager would ask the Executive Committee for permission to approach that person directly. The AED was looking for engineers, preferably young ones, who had worked for DuPont for at least one year. To have worked at Hanford was obviously ideal, and Hanford veterans were the first to be asked. Dale F. Babcock, for instance, had been hired by DuPont in 1929 after he received his Ph.D. in chemical engineering from the University of Illinois, where he had studied under Carl "Speed" Marvel. A member of the Nylon Department until 1942, Babcock joined TNX that year and worked there with Worthington on the design of the Hanford reactors. In 1945, Babcock was assigned to the Graselli Chemicals Department, where he remained until 1950, when he was put in charge of the Savannah reactors.[84] Babcock's career constantly moved back and forth between DuPont's military and civilian activities.

The letters offering jobs in the AED stressed the "exciting" nature of an adventure at the leading edge of technology, DuPont's long-term commitment to the nuclear field, and the fact that a great many young men, "like yourself, dear friend," would be working there.[85] The letter also said that the exact nature of the work could not be divulged for security reasons. But that it would surely be "fascinating," and pay the same salary, plus a raise of 10 percent to compensate for the cost of relocating to South Carolina and for shouldering increased responsibility.

What were the results of this policy of transferring engineers and technicians from civilian departments to DuPont's atomic division? The numerical survey indicates that the initial goals were met and sometimes even exceeded, despite a rather slow start. For the year 1950, the goal was to transfer 179 engineers and technicians. By the end of the year, 257 had been brought in. The transfer procedure took time, since every new AED employee had to get a security clearance from the FBI, which took about three months.[86] The quotas were also attained in 1951. In subsequent years, the policy of quotas was abandoned in favor of individual transfers negotiated case by case, and with a view to specific positions, among the AED, the prospective candidate, and his present department.

The AED also recruited employees from outside the company. DuPont's Personnel Department steered certain candidates toward the recruiters of the AED, who then conducted the interviews. The AED also recommended certain applicants with Ph.D.'s, who might have worked for that division as students or postdocs.[87] By 1 July 1952, 1,327 persons from outside the company had been interviewed, 1,145 had received an offer, and 560 had joined the AED.

DuPont's atomic division thus had little difficulty recruiting a sufficient number of employees. Attractive salaries, the perspective of working at the leading edge of technology—the undeniably ambiguous image of the nuclear enterprise had not yet been supplanted by the conquest of space in the American popular imagination—and the possibility of speeding up one's career all furthered the goals of DuPont's directors and the AED.

In addition to the AED employees, the survey also counted those of the construction division of the Department of Engineering. During the five years of construction, it listed between 35,000 and 38,500 persons, 80 percent of whom were directly employed by DuPont, while the remaining 20 percent were employees of subcontractors. By 1951, the Savannah site was the largest employer in South Carolina, and it remained so until its dismantling in the early 1990s.

The work week, initially 45 hours, was increased to 54 hours in March 1952, or six 9-hour days. This was a work rhythm comparable to that of World War II, but it was compensated with substantial overtime bonuses (14 hours per week). The 45-hour week was reinstated only in March 1954. In addition, all new employees had to participate in a series of training sessions in order to familiarize themselves with the project's working methods and safety procedures.

Unlike at Hanford, which had become "Government City," the AEC had this time refused to pay for housing. Nor did DuPont feel that these costs were its responsibility. As a result, some 40 percent of the employees were housed for several years in rudimentary quarters, usually construction shacks, which became overheated in the summer heat. Families were entitled to a shack; unmarried people were housed in dormitories, although many construction workers were from the area and could go home at night. These rustic living conditions were not very attractive, and the company soon had to promise the employees paved streets and air conditioning.[88]

In a widely discussed book on the permanent war economy, the economist Seymour Melman has analyzed the existence of a defense sector cut off from economic realities and exclusively dependent on military contracts.[89] More recently, the geographer Ann Markusen has analyzed the military geography of American industry in terms of a "gunbelt," running along the country's Atlantic and Pacific coasts and through the South, that is generally characterized by a pronounced military bent, an ununionized labor force, and a market disconnected from the civilian world.[90] Useful as these analyses are, one must also consider another facet of the war economy, which overlaps with civilian activities to a much greater extent. DuPont's AED did not have its own personnel and its own set of recruiting procedures different from those of the rest of the company. Many of the engineers only stayed at the Savannah plant for a few years and were then transferred to another of the firm's departments. To use the official terminology, this was a matter of "cross-fertilization" among departments, which called for regular exchanges of personnel. This means that the AED, though commercially cut off from the rest of the company, was not isolated. The military-industrial complex must not be thought of as an airtight network, because to some extent it overlapped not only with the civilian economy but with the very structure of the companies involved. The engineers employed did not seem to experience any ethical concerns: they worked on the production of plutonium and tritium without any discernible moral qualms, just as they had worked on that of nylon and dacron.

The Manufacture of Fissile Materials

Unlike the atomic bomb, which is based on the principle of the fission of uranium atoms, the hydrogen bomb operates by fusion. The principle of fusion was already known before the war.[91] The chemist Harold Urey had received the 1934 Nobel Prize in chemistry for having demonstrated the existence of three hydrogen isotopes and formulated the possibility of atomic fusion.[92] But at the time, the problem that seemed insoluble to the physicists was that to bring about the fusion of atoms, an enormous amount of energy is needed, an amount that must be greater than the energy obtained from their fusion, since the nuclei have positive and negative charges, which repel each other. But then the atom bomb developed during the war shed new light on the problem of fusion, for it was learned that during nuclear fission a temperature of several million degrees was reached in a few millionths of a second, and that neutrons were projected in every direction. As early as 1944, Edward Teller had the idea that an atomic bomb could be used to trigger a hydrogen bomb.

The principle of the hydrogen bomb, then, is the following: a core of ^{235}uranium or plutonium is surrounded with tritium, deuterium, or lithium (three protons and four neutrons which, when combined with a proton, yield two helium atoms). When the uranium splits, neutrons projected at high speed fuse with the tritium or the deuterium. The more tritium or deuterium is present, the more powerful the bomb becomes, a feature that makes for unlimited possibilities (whereas in an atomic bomb, the quantity of uranium is limited, for when a certain mass, the so-called critical mass, is reached, fission occurs spontaneously and the bomb misfires.)

To be sure, this theoretical principle had yet to be applied. And it is here that difficulties arose. In the spring of 1950, the calculations of the mathematician Stanislaw Ulam showed that, contrary to Teller's optimistic predictions, several kilograms of tritium would be needed for each hydrogen bomb, and that even then fusion was not guaranteed.[93] The final design of the bomb was not worked out until March 1951, when Ulam and Teller had the idea of using the radiation emitted by the fission a few millionths of a second before the explosion to compress the deuterium, thereby making the fusion possible.[94] But at the time when DuPont returned to an active role in the nuclear field, no one knew exactly whether the thermonuclear bomb could actually be built.

The Savannah River facilities therefore had to be sufficiently flexible to produce a large range of fissile materials. Hence the technical choice was different here than at Hanford: instead of being cooled by water and using graphite as moderator, the Savannah reactors were cooled and moderated by heavy water, which has the double advantage of having the same thermic qualities as ordinary water and slowing down the neutrons, thereby allowing the chain reaction to take place. The layout consisted of a series of uranium rods submerged in vats of heavy water, which were themselves cooled by water from the river. The main advantage of this process was that the rods could be more easily manipulated than in a graphite reactor; unlike reactors for civilian use, whose objective is to produce heat that generators will transform into electricity, reactors for military use have the objective of producing highly radioactive materials. In civilian reactors, the uranium rods are replaced rather infrequently (perhaps every two years). In military reactors, they are constantly manipulated.[95]

One major difficulty was to find a way to produce heavy water in very large quantities, since each reactor needed 250 tons of it, and the total American stock of heavy water amounted to no more than 25 tons in 1950. Moreover, heavy water becomes extremely hot and highly radioactive upon contact with uranium and therefore had to be contained in watertight vats, which raised additional problems of design. In fact, this was the very reason why during the war the engineers and physicists, being pressed for time, had opted for reactors filled with graphite. A first phase therefore had to be devoted to building plants that used the water of the Savannah River to produce heavy water by means of repeated distillations.[96]

In a second phase the engineers went to work on the reactors themselves. These were vats about five meters in diameter filled with heavy water and loaded vertically with uranium rods and control rods. Around the reactors a strong flow of ordinary water that cooled the heavy water was itself cooled down as it flowed back and forth between the vat and a set of heat exchangers. The danger of contamination was fairly high, given the intense radioactivity of the 250 tons of heavy water that circulated in the reactor without a completely watertight barrier to separate it from the ordinary water and the surrounding air. Although the precautions taken here were more stringent than at Hanford, the reactors' high chimneys allowed radioactive gases to escape, hydrogen sulfite in particular, while the ordinary water, also contaminated, seeped into the clayey and lateritic soil of South Carolina. Twelve huge underground reservoirs were filling up with radioactive materials, heavy water and uranium nitrate, awaiting the day when they could be properly disposed of. This was a time when hardly anybody worried about the danger of pollution.

In a third phase, the engineers had to proceed to the chemical separation of the irradiated elements, as had been done at Hanford. Operations at the two large plants built for this purpose were carried out from a distance by remote-controlled cranes. The methods used here to extract the plutonium and tritium obtained by the prolonged irradiation of ^{235}uranium were more effective than at Hanford.

The first reactor was put into operation on 28 December 1953, followed by the four others in the course of 1954 and in early 1955. The first load of enriched uranium was extracted in June 1954 and separated over the following weeks in order to obtain plutonium.[97] The first fissile materials, plutonium and tritium, were delivered to the AEC on 29 December 1954, more than two years after the first hydrogen bomb, dubbed "Mike," was detonated on 1 November 1952. Savannah was ready to bring the United States into the era of mass-produced nuclear bombs. Beginning in late 1956, the five reactors were operating continually and at full capacity.[98] Until they were dismantled, they constituted the only source of tritium and the major source of heavy water in the United States, and both were needed for making hydrogen bombs, which were produced at the rate of several hundred per month, until their number reached the record high of 32,500 in 1967.[99]

The building and management of the Savannah complex reinforced the ties between DuPont and the AEC. Aside from their close cooperation at the site itself, the capital of expertise that DuPont had accumulated earlier was extensively drawn upon by the AEC and its successor agencies.[100] Several engineers, among them Babcock and Evans, sat on the General Advisory Committee to the commission, following in the footsteps of their colleague Worthington. One DuPont engineer was also lent to the Washington, D.C., office of the AEC, and later, in 1975, to the U.S. Department of Energy.[101] Even though DuPont derived no financial benefits from its involvement in the nuclear field, its participation was a way to keep in touch with a potentially lucrative body of knowledge, a training school for the company's engineers and chemists, and also a source of prestige in an era when nuclear activities had not yet been supplanted by the conquest of space and still represented the last frontier of technology.

In October 1961, Greenewalt took a revealing initiative. In a letter to Glenn Seaborg, president of the AEC, the president of DuPont took stock of his company's nuclear experience:

Dear Glen,

You will recall that when Du Pont left Hanford, and again when the Savannah River project was started, brief statements were made by us to the effect that these jobs were undertaken solely at the request of the Government, and that the Company disclaimed any commercial interest in the nuclear field. These statements were made in good faith, as you know. . . . More than ten years have passed, and they have witnessed to a dramatic extent the overlapping and diffusion of scientific disciplines and fields of technology. This is certainly true in the chemical industry, where lines of demarcation are far less sharp than in the pre-Savannah days. The nuclear field has also broadened so that there are now increasingly greater areas of overlapping interest.

This is another way of saying that the reasons for Du Pont's past reluctance to become more deeply involved in the nuclear energy field are no longer entirely applicable. More specifically, we are willing to undertake further extension of work under the Savannah River contract to include production of other materials which the Commission is called upon to produce or any mutually agreeable research and development programs in the nuclear field. We are also interested in investigating appropriate areas outside the contract which may be commercially attractive to us.

We recognize [that], as a result of past history, the general impression exists that Du Pont is not a competitor in the nuclear field except as a supplier of our commercial products. Obviously, we expect to take steps to correct this impression.

Under the circumstances, before taking any such steps, we believe we have an obligation to advise the Commission of our present interest. We do not have any specific projects to propose at the present time but we are hopeful that some may develop. We will be very happy to explore with the Commission any areas that may seem promising to it, and we shall of course bring to your attention those that occur to us.

Sincerely,

C. H. Greenewalt

President[102]

Having extended its five-year contract with the AEC, even though it was under no obligation to do so, and even though the worst of the Cold War was over, DuPont now hoped to extend its activities into civilian pursuits in cooperation with the AEC. This was already beginning to be done on a modest scale, for since 1956, under the aegis of the AEC, Savannah had been producing a number of weakly radioactive items for civilian use, such as ^{252}californium for the treatment of cancer. But Greenewalt wanted more and envisaged DuPont becoming—and why not?—a contractor for electricity production.[103]

To be sure, in 1961 all the big chemical and electric companies thought that nuclear energy for civilian use would become a major source of profit in the near future, and DuPont, like all the others, did not want to miss the boat. Moreover, one must also take into consideration the personal choices of Greenewalt, who had witnessed the first chain reaction in history under the stands of the stadium of the University of Chicago in December 1942 and had always been very interested in this sector. That DuPont in the end did not become a major operator in the field of civilian nuclear activity is beside the point here: Greenewalt's letter of intent is valuable for what it reveals about the mutation his company had undergone over twenty years. It is true that DuPont did not go as far as General Electric, McDonnell Douglas, or Lockheed in its collaboration with the government, for it turned down several major military contracts for the development and production of missiles and torpedoes and carried out its military nuclear activities without financial gain. But the "industrial" side of what Eisenhower in his farewell address of 1961 called the "military-industrial complex" did not consist only of a few major contractors working exclusively for the military. It included large companies essentially engaged in civilian pursuits. One of these was DuPont, whose political past had not exactly foreshadowed taking on heavy commitments such as Hanford and Savannah either.

For its part, the professional association of chemical engineers, the AIChE, also encouraged the chemical industry to accept government contracts. In an editorial in *Chemical Engineering Progress*, R. C. Gunness declared that the imperatives of national defense had to be the first consideration, and that, furthermore, once the emergency was over, industry would be in a good position to hire large numbers of engineers. And when all was said and done, Gunness added, industry did a better job than the government. And although government contracts might be a headache for the administrator, the results were favorable in terms of training personnel, national defense, and the well-being of the general public.[104] For different reasons, both DuPont and the AIChE found the institutional arrangement that had been concluded between the chemical company and the Atomic Energy Commission to be in their interests.

The Golden Age of the Chemical Engineers

Basking in the glory of their achievements, the chemical engineers of the 1950s enjoyed a favorable image. The surveys conducted by *Chemical Engineering Progress* report that many students were attracted to chemical engineering, which ranked second after electrical engineering, a field that, increasingly moving toward electronics, fascinated young American engineering students more than others. The proportion of chemical engineers in the total number of engineers climbed slightly (from 9.8% in 1946 to 11.7% in 1950 and 12.1% in 1955). Whereas mining engineers were fewer in number in both absolute and relative terms, the proportions of civil and mechanical engineers remained stable, and the cohorts of electric and electronic engineers were growing larger. The two new specialties of the twentieth century thus accentuated their relative advance over the others.

In 1955, DuPont officially counted 3,403 chemical engineers, an increase of 118 percent since the war, and also about 10 percent of all chemical engineers in the United States (34,090 in 1950). The newcomers knew about the accomplishments of their predecessors: they had been told that DuPont engineers were men for whom nothing was impossible, who were capable of great exploits. The company brochure almost made the chemical engineer into a superman; one sketch, for instance, represented him as a Gulliver hovering over a laboratory and a factory. Indeed, this may have been overdone. A note from Laid Stabler of the Department of Human Resources made the point that everyone had his place at DuPont, including those who were not destined to reach the top.[105] This manager seemed to fear that the prestige of the great elders might discourage employees with lesser ambitions and resources.[106]

A more general statistical picture can be drawn on the basis of the information collected by the Department of Labor in 1953 in the first official study devoted specifically to the chemical engineers.[107] In late 1951, the investigator, Laure Sharp, sent out questionnaires and received responses from 13,400 chemical engineers, or 39 percent of the total accounted for in the census of 1950.[108] Her first finding was the almost total absence of women. There were only 37 of them (or 0.1%, compared to 7% of the chemists). This figure confirms what we know about DuPont. Moreover, 96 percent of the responding engineers had been born in the United States, as against 3 percent who were naturalized citizens and 1 percent of foreign nationality.

A third finding was that chemical engineers were one of the youngest professional groups in the United States: their average age was thirty-two. Two in three were under thirty-five, and 80 percent were under forty.[109] Chemical engineering was the youngest engineering profession; mechanical engineering, where the average age was thirty-six, was the second youngest, and the civil engineers were the oldest, with an average age of forty-five. The overwhelming majority of the chemical engineers had graduated in the late 1930s and in the 1940s and 1950s.

Although the number of postgraduate degrees showed a steep rise in the postwar years, most engineers had had only a bachelor's degree in engineering. Seventy-one percent of the chemical engineers had a bachelor's degree, 20 percent a master's degree, and 8 percent a doctorate. The study also found 2 percent of engineers without degrees—though with some college training. At an average age of forty-eight, this last group was the eldest in the sample, which means that it represented the first generation of engineers, men like Robert MacMullin, who had been trained on the job in the late 1910s. Noteworthy too was the fact that 97 percent of the chemical engineers had at least a bachelor's degree in chemical engineering, whereas a fourth of the chemists came from another field, usually mechanical engineering.[110]

The employment pattern of these engineers is instructive as well. Eighty-four percent of them worked in private industry (with the remaining 16% equally divided among the federal and state governments, teaching institutions, and engineering consulting firms). Most of them were employed either in the chemical (48.3%) or the petrochemical (21.2%) industries; the others were scattered among the different officially recognized manufacturing industries. The survey also made it clear that the number of chemical engineers engaged in research and development was much higher than in other engineering specialties: 19 percent stated that research was their principal activity (compared to an average of 7% for other engineers), and for 17 percent, it was development (compared to 9% elsewhere). We thus arrive at the astonishing figure of 36 percent of chemical engineers engaged in research and development. This proportion should not necessarily be taken at face value, for it is possible that the engineers who responded were those who had the most prestigious careers or were working for large companies. Nonetheless, it does indicate a clear upward trend of the profession. Working in production came in second place, with 28 percent, followed by design (12%) and management (10%).[111] This last and relatively low percentage is explained by the youth of the chemical engineers, for it is a fact that managerial careers usually took shape after a certain number of years in research and development or production. Three-fourths of engineer-managers were over thirty-five.[112]

The salaries of American chemical engineers in the 1950s reflected their enviable socioprofessional position. The average salary was $5,600, but this average hides disparities related to seniority, credentials, and type of employer.[113] It is not surprising that a young engineer with a bachelor's degree was hired at $3,700, as compared to $11,700 or more for an engineer at the end of his career (aged 60–64), whose salary had risen particularly rapidly if he was employed in the chemical or petrochemical industries.[114] Such levels of remuneration made the chemical engineers the best-paid group of engineers in the United States, with the electrical engineers following closely behind, with an average of $5,500. By contrast, chemists employed in industry, a group that had dominated industrial chemistry a half century earlier, were much less well paid, receiving $3,400 rather than $3,700 as a starting salary; $6,500 rather than $7,300 at age 35–39; and $7,800 rather than $11,000 at age 50–54.[115]

The profession as a whole was thus rather homogeneous in demographic and cultural terms: these were middle-class young white men, whom company magazines and yearbook photos show dressed identically in dark trousers, white shirts, dark neckties. Sometimes demonstratively rolled-up sleeves, a discreetly open collar, or a loosened necktie symbolically convey the idea that the engineer is a white-collar person who is not afraid to dirty his hands.

Their political preferences were rather homogeneous as well. According to a survey conducted by *Chemical Engineering Progress* just before the presidential election of 1952, 60 percent of the chemical engineers identified themselves as Republicans, compared to 39 percent of college graduates as a whole; 12 percent as Democrats (as against 26%), and 28 percent as independents (as against 35%). Chemical engineers were one of the most conservative groups, the survey concluded.[116] From a sociological point of view, they showed themselves to be rather satisfied with their professional lives and quite willing to adopt the political and social values that prevailed in the upper echelons of their organizations.

Back to Academia

Aside from the number of chemical engineers or students in that field, however, what is particularly remarkable in this postwar era is the lengthening of the training in chemical engineering, which often went to a master's degree, and sometimes to a doctorate. There was a definite increase in doctoral programs in chemical engineering. American universities produced 50 to 60 doctors of chemical engineering annually in the 1930s and 1940s, and about 200 in the 1950s. This figure climbed regularly until 1972, the record year (453 doctorates in chemical engineering), then decreased in the 1970s until it stabilized around 200 again in the 1980s.[117]

The strong postwar increase in doctorates of chemical engineering had to do, first of all, with the incorporation of applied mathematics and thermodynamics into the field. We have looked at the development of "transport phenomena" in the 1930s in connection with the advances in high-pressure chemistry. Transport phenomena became extremely important after the war, given the great demand caused by the parallel and spectacular rise of the synthetic fiber industry, military and civilian nuclear endeavors, and petrochemistry. All of these fields involved complex phenomena of thermodynamics, whether in manufacturing polymers or cooling nuclear reactors. In every case, the mathematical approach proved useful. Unit operations had faded into the background, making room for a more dynamic conception of chemical processes. More than a third of doctoral dissertations in the field were related to unit operations in the 1930s, as were half of them in the 1940s, but in the 1950s, this proportion declined and settled at about 30 percent. In that decade, transport phenomena were the subject of almost 55 percent of dissertations.[118] The complexity of these physico-chemical phenomena often motivated students given to mathematical formalization to prolong their studies until they were given a thesis topic.

The growth of hard science in the training for chemical engineering was not, of course, unique to the world of the American engineers. Electrical engineering also incorporated more physics and mathematics, leaving problems of electricity behind and taking an increased interest in electronics, computer science, and solid-state physics in general. Electrical and chemical engineering now tended to become more elitist fields in their emphasis on science, thereby responding to the growing demand for experts voiced by major industries engaged in either military or civilian pursuits. After the launching of Sputnik, the scientific programs of the engineering schools were strengthened even further in response to the conclusions of a federal commission that had expressed concern that American engineers were lagging behind their Soviet competition in certain scientific areas.[119]

In addition, departments of chemical engineering now had more funding for financing their doctoral students than in the past. One professor of chemical engineering at the University of Pennsylvania remembered his department's flourishing finances in the 1950s.[120] Chemical companies were providing many more fellowships annually, and all the students had financial support. DuPont as well wanted to do its part and distributed a wide variety of fellowships. In 1952, it granted 1,037 fellowships, at a cost of $275,000.[121] It also hired a larger number of doctors of engineering, which prompted Greenewalt to confide that he would not have been hired if he had knocked at the company's door with only his bachelor's degree after the war.[122]

DuPont strengthened its collaboration with the universities in 1951, when it launched a "Year in Industry Program" that allowed one or two professors of chemical or mechanical engineering to spend a year in the firm. Not only were they free to observe and to become involved in specific industrial operations or research programs, but they also participated in meetings, attended training sessions in different plants, and could even, if they so desired, take charge of a specific project.[123] Bart Conta, for instance, a professor of chemical engineering at Cornell, carried out calculations for heat transfers in connection with a new cellophane plant—a task of particular interest to him since he was writing a textbook of thermodynamics.[124]

Once they had returned to their university departments, these professors incorporated their industrial experience into their teaching programs. The interpenetration between academic departments and the chemical firm, an important phenomenon even before the war, now served to bring more hard science into the discipline. Henceforth the professional identity of the chemical engineer was attuned to a military and civilian demand that required a certain type of training and specific technological qualifications. The attraction of industry was such that DuPont was concerned that a shortage of Ph.D.-holding professors of chemical engineering might ensue. The "Year-in-Industry Program" was designed precisely to keep the professors happy, as was the creation of endowed chairs. In 1957, DuPont, and Greenewalt personally, thus endowed the Warren K. Lewis Chair in Chemical Engineering at MIT, honoring the celebrated author of *Principles of Chemical Engineering;* the Allan P. Colburn Chair at the University of Delaware was endowed a year later.

Another major characteristic of chemical engineering in its postwar era of triumph was its close involvement with the nuclear field. DuPont's participation in the Manhattan Project and then in the thermonuclear program had opened up new approaches that academic departments of chemical engineering were eager to pursue. This field provided them with the opportunity to enlarge their students' career choices, to develop new programs, and to find new financing. As early as 1946, MIT's department of chemical engineering had created a division of "nuclear engineering." Manson Benedict was appointed professor of nuclear engineering in 1951. Over the next few years, three more professors—Robert Pigford, formerly of DuPont's Department of Engineering, Edward Mason, and Theos Thompson—joined him. A small nuclear reactor designed by Thompson was set up on the MIT campus in cooperation with neighboring hospitals. In those years, the development of nuclear engineering was such that it was only logical for this division of MIT to assume its independence and to become the autonomous department of Nuclear Engineering in 1958.[125] At most other major universities, programs in nuclear engineering were set up directly by departments of chemical engineering.

Any worries the chemical engineers might have were those of the spoiled children of the American century. *Chemical Engineering Progress* devoted many articles to the promotion of engineers and to the rapidity with which they migrated into managerial ranks, comparing chemical engineers to their colleagues in electrical and mechanical engineering. Increasingly, the position of engineer was considered a stepping-stone to management: According to *Chemical Engineering Progress,* the engineering work should ideally last about ten years. This makes engineering an unusual profession, since physicians, academics, and lawyers normally practice theirs throughout their working lives. Actually, this evolution of the engineering career was not new: even before World War I, the incipient bureaucracies of large corporations had employed engineer-managers, and the du Pont brothers had promoted a generation of young engineers to higher positions in the years before 1920.[126] But in the course of the century, the increasing complexity of the management systems in large corporations placed engineers in an enviable position in the race for the best managerial positions. There are no statistics on the evolution of engineering careers at DuPont. However, the engineers whose careers we have followed here all ended up in management positions. By the mid 1950s, DuPont engineers born in the early years of the century who had earned their stripes in the development of nylon and plutonium—and in some cases in their role as go-betweens in the new dialogue between their company and the federal government—were at the height of their careers. Significantly, the AIChE, which counted 18,000 members in 1958, compared to 872 in 1930, was now admitted to the United Engineering Trustees, a body founded in 1903 at Andrew Carnegie's initiative to bring together all the engineers in the United States.

The Temporary Triumph of Expertise

The years between the end of the war and the 1960s truly were the golden age of American engineering, particularly for those who specialized in chemical and electrical engineering. Highly favorable representations singled them out as indispensable experts who were needed to safeguard the country's security and prosperity. In April 1951, *Time* devoted a long article to DuPont, with considerable attention given to its chairman, who was even depicted on the cover, against a background of a forest of columns, pipes, test tubes, stills, and two bombs. The article celebrated the "full maturity" of the company in terms reminiscent of those used by Lilienthal in *Big Business,* and appropriately pointed out that DuPont was "a company of engineers, run by engineers."[127]

But more was involved here than a discourse on the benefits to be derived from technology and large corporations. What was also being promoted was a certain conception of politics, of the meaning and the effectiveness of political action, which assigned a central position to the neutral and positive figure of the expert.

This conception was not entirely new. Early in the century, Frederick Taylor's scientific management had not presented itself as a mere technique of production, but also as a "true science" that promoted the effectiveness and the rationality of the engineer, in contrast with the sterile jousting of the politicians.[128] Herbert Hoover, who before his tenure in the White House had been the charismatic president of the Association of Mechanical Engineers, had also urged engineers to instill certain principles of their technical rationality into politics. Some essayists, however, went beyond Hoover's modest proposals. Among these were Arthur Morgan, the first president of the TVA, who saw the engineer as an emblematic figure of political modernity, and particularly Thorstein Veblen, who felt that engineers should occupy a central position in the new society he envisaged.[129] Howard Scott's Technocratic Movement, which had its hour of glory during the dark depression years, advocated a social reorganization that gave "experts in science and technology" a central place in politics.[130] And although the New Deal did not grant engineers a particularly eminent position, it did value social engineering when it created the Tennessee Valley Authority and placed it under the command of the engineer Morgan and the lawyer Lilienthal.[131]

But the postwar period marked a true turning point, for America's technical systems had reached a new qualitative stage in the 1940s, when they branched out and when the research and development operations of corporations melded with those of the government. A political context that favored large organizations afforded engineers and administrators the historic opportunity to form a solid partnership, what Brian Balogh has called the "prominentistrative State" composed of professionals and administrators.[132]

Many politicians and business leaders came to feel that the political oppositions of the past no longer made sense. This is what John F. Kennedy expressed in his own way in his commencement address at Yale University on 11 June 1962:

> It is not new that past debates should obscure present realities. But the damage of such a false dialogue is greater today than ever before simply because today the safety of all the world—the very future of freedom—depends as never before on the sensible and clearheaded management of the domestic affairs of the United States.
>
> . . . And we cannot understand and attack our contemporary problems in 1962 if we are bound by traditional labels and worn out slogans of an earlier era. But the unfortunate fact of the matter is that our rhetoric has not kept pace with the speed of social and economic change. Our political debates, our public discourse—on current domestic and economic issues—too often bear little or no relation to the actual problems the United States faces.
>
> What is at stake in our economic decisions today is not some grand warfare of rival ideologies which will sweep the country with passion, but the practical management of a modern economy. What we need is not labels and cliches but more basic discussion of the sophisticated and technical questions involved in keeping a great economic machinery moving ahead.[133]

Among those who concretely exemplified this new attitude were the engineers working at nuclear sites or in nylon plants. Greenewalt himself understood this quite well, as he showed when he addressed an international meeting of industrialists in terms that would have sounded strange indeed to the ears of a Pierre du Pont in the 1920s:

> [B]usiness leadership must take an active and constructive position toward the functioning of society as a whole. All too often, I am afraid, we in the business community have been remiss in this respect. In our attitude toward government, for example, we have been inclined to take a negative position, characterized more by opposition to social change than by a thoughtful and cooperative con

sideration of broad national problems. Too often have business people complained about what "those fellows" in Washington, in Paris, in Bonn or in Tokyo were doing to them, rather than concerning themselves with whether what was being done was right. . . .

If the growth of the corporation has bred a professional class of business managers, the growth of government seems to be breeding a professional class of administrators. All too often, I am afraid, we have seen the two arrayed one against the other in attitudes of mutual distrust and suspicion. . . .

Today, in sharp contrast, both business and government draw their leaders from the same pool of talent, and the educational backgrounds of the two, allowing for scientific specialization, are not dissimilar.[134]

To be sure, it would be inappropriate to consider the chemical engineer as the deus ex machina of the era of consensus and of the "end of ideology." When Kennedy and Greenewalt spoke about the era of the expert, they were not specifically, or exclusively, thinking of the engineers. Yet the fact is that engineers were among the most industrious promoters of the postwar encounter between American managers and bureaucrats. This encounter became the symbol of innovation and efficiency: the threat of domestic social unrest could be defused by economic growth, and the Soviet menace could be met by combining the military arsenal with technical know-how.

American historiography reflected this mutation. In this respect, the historian Richard Hofstadter is emblematic of a turn in the writing of history. His first book, *Social Darwinism in American Thought,* published in 1944, agreed with Charles Beard's theses concerning the dangers of monopolies and internationalism and warned Americans against a conflict that would only reinforce the positions of capitalism and weaken democracy.[135] When World War II proved the Progressive historians completely wrong, Hofstadter felt compelled to acknowledge this mutation. Giving up the binary opposition between democracy (the people, the American virtues) and capitalism (the special interests, the European influences), he published *The American Political Tradition* in 1948 and *The Age of Reform* in 1955, two works in which he judged the Progressives severely. For him, the danger did not come so much from big business as from the old-fashioned populist and Progressive leaders: "They had been brought up to think of the well-being of society not merely in structural terms—not as something resting upon the sums of its technique and efficiency—but in moral terms, as a reward for the sum total of individual qualities and personal merits."[136]

Hofstadter predicted that the Progressive legacy was about to be irremediably swept away by the corporations and by the exigencies of foreign policy. Some of Hofstadter's analyses of populism and Progressivism have been questioned since then, but what interests us here is that this great historian contrasted the "Age of Reform"—which had ended with the New Deal—with the "Age of Consensus," which had begun with the war. Unmoved by nostalgia, Hofstadter welcomed the age of consensus, and also the abandonment of isolationism. He now hoped for the unfolding of a new society that had broken once and for all with the populist tradition and with the irrational crusades the Progressives had led against the great capitalist organizations. According to Hofstadter, the new heroes would be the experts, who were capable of managing the economy and society without partisan passion, in a spirit of cooperation between the public and the private sphere. In the same vein, he called on his fellow historians to let go of the moralizing judgments of the Progressive historians and to turn instead to dispassionate analyses. Above all, he told them, they should be aware of the irreversible nature of the development of the great organizations and avoid depicting them a priori as the adversaries of the general interest.

This is not to affirm an idyllic and purely functionalist vision of a postwar America miraculously purged of all social conflict. During the war, there had been 14,471 strikes, most of them "wildcat" strikes.[137] Despite the enactment of the Taft-Hartley Act, there were even more strikes during the Korean War, and these involved more workers (92.6 million) than those of the period 1935–39 or World War II.[138] But the turn toward conservatism that the American unions had taken in the late 1940s made it clear that labor had been captured by the consumer ideals of the middle classes.

Moreover, part of the American population, notably the African Americans, still did not partake of this prosperity. But the corporations were increasingly interested in the black consumers, especially the lower middle classes of the northern cities. The first DuPont advertisements to target them specifically—they were for DuPont's plastic housewares—appeared in the spring of 1949.[139] Black and working-class people constituted a market that the large corporations coveted.

Meanwhile, in the early 1950s, the majority of Americans, while recognizing the existence of deeply rooted political and social problems, still believed that they had found in the mass production of sophisticated consumer goods under the apparently efficient and neutral aegis of engineers and managers the best means of maintaining regular growth, defusing social tensions, and fostering reconciliation in a pacified society. Less than thirty years after the founding of the Liberty League, Greenewalt expressed his gratification at the rapprochement between managers and federal officials—in other words, at the coming to power of a class of experts that he and his friend Lilienthal embodied to perfection. This class was determined to go beyond some of the major political polarities of the prewar era and to replace them with the trenchant rationality of technical expertise. In a sense, some of the historical interest of the American nuclear program and the development of nylon has to do with the fact that they were for a time presented as test beds of managerial, commercial, and political modernity. It would appear that more studies are needed if we are to analyze these moments when politics was eclipsed, obscured by a purely functional rationality and the triumph of what Dominique Janicaud has called the "techno-discourse."[140] But they are also needed if we are to analyze the periods of political resurgence that have arisen in the United States, spurred by the recent recognition that growth in technical and organizational power does not automatically mean growth in prosperity and security.

The Waning of the Days of Glory

In the 1960s, DuPont experienced something of a crisis, and this was a shock to the chemical engineering profession, which had been entirely constructed according to the modernist paradigm.

To begin with, European competition, particularly from Germany, again made itself clearly felt. Right after the war, the Americans had had little to fear from a European chemical industry that lay in ruins. The dismantling of IG Farben, which restored the autonomy of Bayer, BASF, and Hoechst, did not hold the Germans back for long, and they soon became leaders once again in their special fields, the dyes and pharmaceutical markets, American customs barriers notwithstanding. Moreover, in the 1970s, German companies began to invest directly in the United States by setting up branches and entering into partnerships with small American firms, while in Europe itself, the Common Market favored the growth of French (Rhône-Poulenc) and Italian (Montedison) firms. Competition within the United States intensified as well. The other major chemical companies (Monsanto, Union Carbide, Dow) could now make life difficult for DuPont, whose technological and commercial lead had melted away. As a journalist writing for *Fortune* pointed out in 1961, DuPont's major competitors had adopted its methods, and aggressive small firms had entered the market by taking over lucrative niches.[141] Moreover, engineering firms could now supply turnkey plants to anyone who wished to branch out into chemicals.

The year 1965 was one record profits for DuPont ($800 million), but the situation deteriorated over the next few years, even though in the aggregate, the company remained in excellent financial health, financing itself entirely out of its own resources, while also paying its shareholders handsome dividends. Risky investments made by Lammot du Pont Copeland, who had succeeded Greenewalt as president in 1962, caused some fragility in the company's finances, but the growth in its sales remained substantial, especially in Europe. The resounding failure of the artificial leather Corfam in particular shook the certainties of the directors, who until then had placed their full confidence in the research model initiated with the nylon campaign. The Corfam venture ended with a net loss of about $80 million. DuPont's fine commercial and technological management machine thus seemed to show some signs of wear. Despite heavy investments, the chemists had not invented the "new nylons" of the directors' dreams. Indeed, by the late 1950s, the most lucid of the managers already had an inkling that there never would be another nylon. Increasing investment in chemical research would not automatically generate revolutionary products. When all was said and done, nylon had been brought about by exceptional circumstances that history would not reproduce.

Furthermore, decentralization also took its toll, for it had created duplication of research and a certain lack of cooperation among departments.[142] But the Executive Committee was divided as to how to remedy this: should the company diversify and invest in new areas, as Greenewalt wished to do?[143] DuPont could call on a considerable reserve of ready money (more than $300 million in 1960), and for a moment considered acquiring an airline or an oil company, and even becoming heavily involved in commercial nuclear energy or space exploration. But on 21 March 1961, a ruling of the Supreme Court ordering the chemical firm to sell its General Motors shares ended its hopes for diversification, at least for the moment. It is clear that DuPont found itself in a strategic impasse from which it emerged only in the 1970s, when it went into biochemistry and pharmaceuticals. In part, the situation was turned around by Charles McCoy, Copeland's successor (1967–73), and Irving Shapiro (1973–80), who ended the strategy of indiscriminate expansion and cut production costs.

Yet even so the era of DuPont's undisputed domination of the world market for chemicals was indeed over, and the company had somehow come down from its pedestal.

To this one must add a social and cultural context that had been undergoing a profound mutation since the middle of the 1960s and had led to a questioning of the authority of experts and elites of every kind. A new concern for the environment came to the fore, fueled by articles in the press and books that denounced the ravages industry had inflicted on the natural environment. Rachel Carson's meticulous exposé in *Silent Spring*, in particular, had a lasting impact on the Americans, who were no longer willing to go along with the modernist credo, according to which the benefits of the consumer society outweighed the temporary problems of pollution. The chemical industry found itself in the first line of fire, particularly since it had been the object of repeated accusations concerning its polluting wastes and the poisoning of living creatures ever since the early years of the twentieth century.

The real issue here was a certain erosion of the image of chemistry, marked by increasing concern about environmental problems and toxicity on the part of the public. Thus in 1962, the year when *Silent Spring* was published, the case of thalidomide, a sleeping pill used to treat morning sickness during pregnancy, which caused birth defects, attracted a great deal of attention in Europe and the United States. Also related to this new attitude was the strengthening of the Federal Hazardous Substances Labeling Act of 1976.

Mechanically linking the growing concern over chemical production with the first signs of disaffection regarding chemical engineering would be too hasty; yet, as Charles Rosenberg has put it, a "background noise" had appeared and was to grow louder in the 1970s and 1980s, when the nuclear industry also came under scrutiny. And it was then that the American public became aware of several alarming reports of the Department of Energy concerning the principal American nuclear sites. At Hanford, the Columbia River had been seriously polluted, especially during the first months of 1945, when significant amounts of plutonium had to be produced under emergency conditions. Of the 200,000 cubic meters of highly radioactive liquid waste, stored in 177 underground reservoirs, 40,000 had leached into the subsoil. At the Savannah River site, the situation was not as serious, but a report that DuPont brought out in 1985 revealed that some thirty "serious incidents" had occurred but been hidden from the public authorities since 1952.[144] Now that the Cold War had gone on for several decades, the Americans had to deal with a considerable mass of radioactive waste. Since 1989, more than $60 billion has been spent for the Department of Energy cleanup at Hanford, and "an additional $200 billion is estimated as needed . . . over the next several decades. Hanford's budget alone is bigger than the Environmental Protection Agency's entire Superfund program," Robert Alvarez writes.[145]

These concerns were given political form by consumer groups who found an eloquent spokesman in Ralph Nader and did not shy away from denouncing the collusion between the chemical industry and certain members of Congress who had been the subjects of intense lobbying efforts. In response to concerted pressure by environmental groups, the federal government created the Environmental Protection Agency in 1972, followed by legislation providing for the control of toxic substances in 1976.

In a more general sense, it was perhaps the modernist credo, which postulated a close association between social progress and the progress of technology, that was breaking down. Had not Americans been told again and again for a century that any social, economic, and military difficulties they might be experiencing could be solved by scientists and engineers? It is true that two world wars had cast a veil of uncertainty over these positivist beliefs, but then the postwar prosperity had revived them. Nylon was the happy symbol of a technological economy working for the well-being of the population.

The cover of *Time*, which had been most laudatory in 1951, was no longer so twenty years later. The powerful political protest movement that had developed in the United States now lumped together large corporations and certain government agencies, accusing them of conducting an imperialist war and devastating the environment. A growing proportion of the population now came to feel that risks were not taken into consideration. The omnipotence of sheer technological efficiency, which subordinated politics to the authority of technical neutrality, threatened to make politics disappear as a public space of liberty.

The obligatory soothing statements on the environment that had been circulated by the chemical industry in the 1960s were now discredited. DuPont's directors were well aware of this fact when they appointed Shapiro—a lawyer rather than an engineer, and also someone who had no family ties to the du Ponts—to the chairmanship. Shapiro was called upon to improve the chemical company's relations with the outside world, in particular, the government and the press.

In short, the image of DuPont—along with that of the chemical industry as a whole—began to deteriorate by the mid 1960s. Chemical engineers recently graduated from the best schools no longer jostled for jobs as they had done a few years earlier. The company's Public Relations Department was sufficiently worried to commission a report about DuPont's reputation among students.[146] This report showed a decline in its reputation since 1965, from 80 to 60 percent of "very" or "somewhat" favorable, and recommended that the company cease centering its communications strategy exclusively on the "past successes" of nylon and the synthetic fibers.[147] The fact was that nylon was going out of style in the 1970s. Once considered noble, synthetics were now seen as tacky. Synthetic fibers had experienced their golden age in 1965, when they accounted for 63 percent of world production of textiles. Now they stagnate around 45 percent. Beginning in the late 1960s, consumers rediscovered the "natural" fibers, rustic, warm living materials: wool and cotton. "When a customer reads 50% nylon, she winces," a retailer asserted.[148] Yet while 64 percent of female French customers prefer natural materials, compared to 30 percent who prefer synthetics, 43 percent choose the latter when shopping.[149] But what can the engineers say in response to the Woolmark ad: "We haven't changed our supplier in 3,000 years"? In this new context, the experts in the mass production of chemicals looked less like emblems of modernity than like its sorcerer's apprentices. Alexander Mackendrick's *The Man in the White Suit*, a famous British film of 1951, is about a chemical engineer, William Stratton, played by Alec Guinness, who invents an extraordinary fabric that does not wear out and does not get dirty; its only shortcoming is that it cannot be dyed. But alas, this pleases neither the textile industry nor the workers, who fear for their jobs. Chased by a mob, the engineer realizes that his famous suit has disintegrated and flees to the sound of jeers. It is telling that while this film was a big hit in Europe in the 1950s, this was not the case in the United States. Could it be that the Americans had learned to see synthetic fibers as perfectly ordinary and devoid of mystery, as the natural meshing of scientific research and mass production, unlike the residents of the Old World, where this concept was more suspect?

And yet a page had been turned for the chemical engineers. The newly negative image of the chemical industry—only 32 percent of Americans considered it favorably in 1971, against 55 percent in 1965—probably played an important role in turning the students against this field. The number of chemical engineering students declined, as did the number of doctorates (from 400 in the early 1970s to fewer than 200 in the 1980s).[150] What is more, many professors felt that the decline was not only quantitative but qualitative as well.

It should be added that the conservatism of the profession, symbolized by its ethnic and cultural homogeneity, was not helpful at a time when demands for integration and civil rights by minorities, particularly African Americans, were having a profound impact. In this respect, DuPont came in for its share of criticism, not only for its clearly discriminatory hiring and promotion policies, but also because Wilmington, the DuPont company town par excellence was—like other American cities—the scene of violent racial conflict in 1967. Lammot du Pont Copeland's brutal and contemptuous response to these events discredited him once and for all.

To be sure, the professional association of the chemical engineers did react to the new times, but in a rather faint-hearted manner. Much remained to be done as far as the minorities and the engineering professions were concerned. Among these, chemical engineering was the specialty with the fewest minority representatives. In the early 1970s, the AIChE established a program of communication and financial aid designed to increase the number of nonwhite engineering students, but the results were disappointing. The local branches were not very cooperative and tended to claim that the potential recipients themselves were not interested in the profession.

The situation has certainly improved since the 1970s, but the number of Hispanic or African American engineers remains very small, despite real efforts by the universities, the government, and certain companies. One of these is DuPont, which since the 1980s has favored recruitment of black and Hispanic chemical engineers. However, beyond the difficulties specific to minorities in an educational system that does not respond well to their needs, the engineering professions present the additional problem of a professional culture built up by and for the white middle classes. This is particularly true of chemical engineering, which has been associated from the outset with a highly capitalistic and socially and culturally rather isolated industry. This profession thus became a bastion of the values that had been built up within big industry since the end of the nineteenth century. In 1997, blacks represented 0.9 percent of the total number of chemical engineers in the United States, as against 2.4 percent in the engineering profession as a whole.[151]

As the engineers of the 1970s looked at their dented status and image, another issue of concern was the environment. Here again they found themselves in the dock, having been too closely associated with a polluting and sometimes arrogant industry. This issue was of particular concern in the universities, where educators began to reconfigure their training programs by incorporating such topics as environmental protection and pollution abatement (the objective of the "Energy and Environment" program of the department of Chemical Engineering at the University of Pennsylvania) into the traditional course work that features transport phenomena and cost analysis. At MIT, two-thirds of the faculty today are working on problems of the environment, pollution abatement, and recycling, as if they had set out to make up for the carelessness of preceding generations.

Conclusion

In the space of a half-century, DuPont's chemical engineers as a professional group achieved an enviable position. Associating the invention and assertion of a new scientific field with the diversification of this company, they also benefited from the parallel rise of the consumer society and the military-industrial complex. To this extent, their success is comparable to that of other professionals, such as electrical engineers, physicians, or lawyers, but what does seem unusual is the fact that this field practically did not exist at all around 1900.

Chemical engineers have been the anonymous and hard-working children of the American century. Certainly, no one has found among them one of those memorable or simply attractive characters whom historians so love to portray. In *The Periodic Table*, Primo Levi confides to Cerrato, one of his former classmates at the Technical Institute of Turin, that he finds it unfair "that the world knows all about the way of life of the physician, the prostitute, the sailor, the assassin, the countess, the ancient Roman, the conspirator, and the Polynesian, and nothing of how we live—we who transmute matter."[1] It is true that in the twentieth century, the chemists' star has faded, outshone by that of the physicists, and that chemists sometimes look like laboratory assistants to the physicists. But this book did not set out to do justice to a group that history had forgotten, and even less to write about a profession simply because this had not been done—which is a rather lazy way to justify historical work.

What really matters is not the historical legacy of the chemical engineers. Instead, it has been my purpose to trace their itinerary in order to show how they were able to make use of the parallel rise of mass consumption and the federal government's power of intervention to further their careers and to fashion a new version of American modernity: sophisticated products for the mass market to ensure the prosperity and the security of the country in a political context that had temporarily set aside the great ideological debates of the past.

What were the principal milestones of their itinerary? The first important stage was the invention of unit operations at MIT, for it provided this field with a conceptual unity by reflecting and reinforcing a new paradigm of cooperation between the academic and the industrial communities. This was not self-evident—the interests of the two parties did not necessarily coincide—but they developed procedures for negotiation, so that by the 1920s, a model for producing specialized knowledge, products, and services was fairly well established in the American chemical industry. Some of its most active proponents were young engineers, who favored formal mathematical procedures that made it possible to solve complex heat-related problems in high-pressure chemistry and also reinforced the identity that set these engineers apart from the grizzled veterans of the powder and explosives plants. Working on the development of ammonia and especially nylon afforded them the opportunity to demonstrate their know-how in close cooperation with the company's chemists. And lastly there was the plutonium venture and the marriage of chemical engineering and nuclear physics, which gave the engineers their patents of political nobility by placing them in strategic positions in the military-industrial complex. Meanwhile, their younger colleagues, recent graduates of colleges and universities, followed in their footsteps when great numbers of them took positions in a rapidly expanding industry. Career choices and technical directions depended, to be sure, on opportunities encountered in a context of constraints and obstacles, but these engineers were highly determined and knew how to evaluate and exploit their possibilities within the framework of two major configurations.

Until World War II, the socioprofessional field in which they operated was that of a company with a clearly marked identity, where the engineers made their way in a relational system codified by the internal rules of the firm, and where career moves were relatively easy to understand. Then the field widened to the point where it included the government, and this made for a much more complex relational system and a multiplicity of new constraints and pitfalls. That is why I refer to this last configuration as political. The traits needed here were no longer those of the past, for in addition to technical competence, it required management and negotiating skills. In a larger sense, the widening of the chemical engineer's field of action went hand in hand with the expansion of his company's relations with other organizations, namely, business institutions (1900s–1920); then institutions of higher learning (beginning in 1920); and finally the federal government (World War I and particularly World War II and its aftermath). The company's structures adapted to the widening of its communications, and the employees best qualified to manage this transformation assumed leadership in the highest echelons of the company. This widening of communications took place under the twofold pressure of the market and the government. An exemplary career was that of Crawford Greenewalt, engineer, then manager, and finally chairman of DuPont, a man who combined technical competence with consummate negotiating skills that allowed him to satisfy the sometimes contradictory wishes of his different interlocutors. Greenewalt knew how to win his case by making sure that he had the support of important partners, and how to negotiate the fine points of a project—in short, he understood the essentially political nature of contemporary big science.

American social history, which began to flourish in the early 1960s, has taken a particular interest in local communities and minority groups, and has also included a multitude of new objects, thus breaking up the field into an infinite number of areas. The concern for in-depth social investigation and micro-historical detail has taken historians away—and this is all to the good—from any temptation to engage in a single imperialist discourse. As Bernard Lepetit has written: "Knowledge of a society is not achieved by reducing it to a single discourse but rather by the rational multiplication of commentaries about it."[2] By showing that in the United States, as everywhere else, history is conditioned by a combination of social processes, and that it has had its share of race and class conflicts, historians have privileged the analysis of differences rather than of consensus. But by the same token, as Olivier Zunz has noted, they have also kept clear of overarching explanations.[3]

Notwithstanding that I have outlined precise areas of research, I feel that we should once again ask larger historical questions. This does not mean, of course, that we should return to the historiographical motives of the 1950s. Many historians agree with this diagnosis, but then it is rather more difficult to achieve such an encounter between the particular and the general, the center and the periphery. In this study I have set out to link three levels of analysis: the level of the professional group, its emergence and its maturation; the middle level of the organization, its technical and political culture; and the macro-historical level of the American political economy and the emergence of the figure of the expert as the central political paradigm of postwar America.

Today the contrast with the years 1900–1960 is striking. Chemical engineering came into its own by playing the modernist card, by betting on the promise of emancipation. Nylon and plastics eventually changed life as it was then known, and in so doing proposed a certain vision of the future, which, based on the idea of technological and social progress, also cast a veil of unconcern over its ecological and human consequences. The atomic bomb itself was a modern object par excellence, in that it claimed to guarantee peace by virtue of its huge power and its ability to make total war impossible. But then, once these modernist illusions had evaporated, the production-centered past of the chemical industry caught up with the engineers, who were henceforth called upon to rectify past mistakes and at the same time to engage in projects compatible with new exigencies. To be sure, fantastic promises of textiles that will adapt to any temperature, come filled to the brim with electronics, or will protect you even in a severe fall still adorn the pages of magazines. But no one believes any longer, no one tries to persuade anyone, that these will change our lives. The end of the modernist paradigm does not condemn chemical engineering but forces it to change, that is, to be aware of a set of complex social and ecological preoccupations that in the past were seen only in terms of profits and losses. We no longer ask engineers to give us the best of all possible worlds, but rather, more modestly, to contribute to a better world. Moreover the Cold War has ended, taking with it a gamut of careers from one end to the other of the military-industrial complex.

The cultural mutations of American society in the late twentieth century undermined the very foundations of the modernizing ideology of the industrial age, which in an overly mechanical manner had believed that high-powered technology, social progress, and the well-being of the population were linked. The most zealous actors in this technical modernity have suffered the backlash against such views, even if they have negotiated a conversion that keeps them working as indispensable practitioners of chemical, pharmaceutical, and biochemical production. In this study I have only sketched certain features of this new era and of this conversion, concentrating instead on the time that came to an end some thirty years ago and was one of the principal facets of the "short" twentieth century: this was the time when the word *nylon* had become a familiar synonym for "perfection" ("It's nylon!"), and when the population seemed to have somehow adjusted to the nuclear threat. Under these circumstances, engineers, and in a larger sense managers and other experts in the large-scale production of consumer goods were no doubt correct when they believed that this was their era, as John F. Kennedy had given them to understand. In 1951, before his success faded into the shadows cast by pressing questions and new concerns about both the excesses of mass production and an unpopular and eventually lost war, Greenewalt could proudly appear on the cover of *Time*. Surrounded by fetish objects of modernity, he is shown in a symbolic array that bears a decided resemblance to certain portraits of sixteenth- and seventeenth-century English or Flemish merchants posing by the side of the pieces of gold and the fine cloth that were the substance of their trade.

Notes

Abbreviations and Acronyms

Acc.   accession number
AIChE   American Institute of Chemical Engineers
HML   Hagley Museum and Library, Wilmington, Delaware
PSDP   Pierre S. du Pont
RG   record group, National Archives

Introduction

1. Frederick Lewis Allen, *The Big Change: America Transforms Itself, 1900–1950* (New York: Harper, 1952), 220.

2. E. I. du Pont de Nemours and Company is referred to for the most part in this book simply as "DuPont," the shorthand form now used by the firm itself. The family, however, writes its name "du Pont."

3. Jacqueline Rémy, *L'Express,* 20 November 1987.

O N E : DuPont and the Rise of Chemical Engineering

1. Martha Moore Trescott, *The Rise of the American Electrochemicals Industry, 1880–1910: Studies in the American Technological Environment* (Westport, Conn.: Greenwood Press, 1981).

2. Charles Reese, "Does Chemical Engineering Pay?" (address to the American Institute of Chemical Engineers and the Chamber of Commerce of Providence, R.I., 24 June 1925), Hagley Museum and Library, Wilmington, Delaware (hereafter cited as HML), Acc. 1706.

3. This information on the American chemical industry is taken from William Haynes, *American Chemical Industry: A History* (6 vols.; New York: Van Nostrand, 1945–54); Ludwig F. Haber, *The Chemical Industry, 1900–1930: International Growth and Technological Change* (Oxford: Clarendon Press, 1971); Bernardette Bensude-Vincent and Isabelle Stengers, *Histoire de la chimie* (Paris: La Découverte, 1993); and Fred Aftalion, *Histoire de la chimie* (Paris: Masson, 1988).

4. Kathryn Steen, "The German Chemical U-Boat," *News from the Beckman Center,* Summer 1993, 6–7.

5. See Thomas P. Hughes, *American Genesis: A Century of Invention and Technological Enthusiasm, 1870–1970* (New York: Viking, 1989), chap. 4.

6. Alfred D. Chandler Jr., *The Visible Hand: The Managerial Revolution in American Business* (Cambridge, Mass.: Belknap Press of Harvard University Press, 1977), 472–74.

7. Pierre Samuel du Pont, the father of Eleuthère Irénée, was elected delegate to the French Constituent Assembly in 1789, representing the town of Nemours. To distinguish himself from another Dupont, Jacques-Charles Dupont de l'Eure, he called himself du Pont de Nemours. A moderate reformer and supporter of a constitutional monarchy, he only escaped the guillotine because of Robespierre's fall. Elected to the Council of Seniors in October 1795, he was thrown into prison again after Napoleon Bonaparte's coup d'état on 9 November 1799 (known in French history as 18 Brumaire). Realizing that he had no chance of a political career in France, Pierre left for America that year, along with his son Eleuthère Irénée, who had worked with the great chemist Antoine Lavoisier, one of the directors of the Royal Administration of Powders and Saltpeters, in homage to whom E. I. du Pont de Nemours and Company was almost called "Lavoisier Mills."

8. See William S. Dutton, *Du Pont: One Hundred and Forty Years* (New York: Scribner, 1942), and id., *Du Pont: The Autobiography of a Scientific Enterprise* (New York: Scribner, 1952).

9. The manner in which the du Pont cousins recovered the firm is analyzed in detail by Alfred D. Chandler Jr. in *Strategy and Structure: Chapters in the History of the American Industrial Enterprise* (Cambridge, Mass.: MIT Press, 1962).

10. Chandler, *Visible Hand*, 487.

11. David A. Hounshell and John Kenly Smith Jr., *Science and Corporate Strategy: Du Pont R&D, 1902–1980* (New York: Cambridge University Press, 1988), 19–26.

12. See Mary-Jo Nye, "Philosophies of Chemistry Since the Eighteenth Century," in Seymour Mauskopf, ed., *Chemical Sciences in the Modern World* (Philadelphia: University of Pennsylvania Press, 1993), 3–24.

13. On chemical engineering in general, the two essential works are William F. Furter, ed., *A Century of Chemical Engineering* (New York: Plenum Press, 1982), and id., ed., *History of Chemical Engineering* (Washington, D.C.: American Chemical Society, 1980).

14. John B. Rae, "Engineers Are People," *Technology and Culture* 16, 3 (June 1975).

15. Paul Starr, *The Social Transformation of American Medicine* (New York: Basic Books, 1982).

16. Terry S. Reynolds, ed., *The Engineer in America: A Historical Anthology from Technology and Culture* (Chicago: University of Chicago Press, 1991), 15.

17. See Bruce Sinclair, "At the Turn of a Screw: William Sellers, the Franklin Institute, and a Standard American Thread," in Reynolds, ed., *Engineer in America*, 151–65; and see also Robert Kanigel, *The One Best Way: Frederick Winslow Taylor and the Enigma of Efficiency* (New York: Viking, 1997).

18. F. W. Taylor, "Why Manufacturers Dislike College Students," cited by Kanigel in *One Best Way*, 138–39.

19. Among these engineers, Major Louis de Tousard, an aide to Lafayette and a renowned artillerist, played a particularly noteworthy role. See Merritt Roe Smith, "Army Ordnance and the 'American System' of Manufacturing, 1815–1861," in id., ed., *Military Enterprise and Technological Change: Perspectives on the American Experience* (Cambridge, Mass.: MIT Press, 1985), 45.

20. Robert F. Hunter, "Turnpike Construction in Antebellum Virginia," *Technology*

and Culture 4, 2 (1963). On the work of the Army engineers, see Forest G. Hill, *Roads, Rails, and Waterways: The Army Engineers and Early Transportation* (Norman: University of Oklahoma Press, 1957), and Daniel Calhoun, *The American Civil Engineer: Origins and Conflict* (Cambridge, Mass.: MIT Press, 1960). See also Françoise Planchot, "Le général Simon Bernard, ingénieur militaire aux États-Unis (1816–31)," *Revue française d'études américaines* 13 (1982): 89–98.

21. Monte A. Calvert, *The Mechanical Engineer in America, 1830–1910: Professional Cultures in Conflict* (Baltimore: Johns Hopkins Press, 1967) 20, 255.

22. On the Morrill Act, see Roger Geiger, *To Advance Knowledge: The Growth of American Research Universities* (New York: Oxford University Press, 1986).

23. See David Noble, *America by Design: Science, Technology and the Rise of Corporate Capitalism* (New York: Knopf, 1977), pt. 1.

24. See Terry S. Reynolds, "The Engineer in Nineteenth-Century America," in id., ed., *Engineer in America*, 25, and Thomas P. Hughes, *Networks of Power: Electrification in Western Society, 1880–1920* (Baltimore: Johns Hopkins University Press, 1983), 363–460.

25. See André Grelon, "Formation et carrière des ingénieurs en France (1880–1939)," in Louis Bergeron and Patrice Bourdelais, eds., *La France n'est-elle pas douée pour l'industrie?* (Paris: Belin, 1998), 231–74.

26. See Chandler, *Visible Hand.* See also Sidney Pollard, *Peaceful Conquest: The Industrialization of Europe, 1760–1970* (New York: Oxford University Press, 1981). Pollard shows that the later a competing company enters the market, the more difficult it will be for it to make use of advanced technology.

27. Edwyn Layton, *The Revolt of the Engineers: Social Responsibility and the Engineering Profession* (Baltimore: Johns Hopkins University Press, 1986), 3. Layton's figures are taken from the censuses. The first American census to use the professional category of engineer was that of 1850.

28. Bruce Seely, "The Scientific Mystique in Engineering: Highway Research at the Bureau of Public Roads, 1918–1940," in Reynolds, ed., *Engineer in America*, 309–42.

29. Olivier Zunz, *Making America Corporate* (Chicago: University of Chicago Press, 1989), 86–87; also Robert Lacey, *Ford: The Men and the Machine* (Boston: Little, Brown, 1986).

30. Robert S. McNamara, with Brian VanDeMark, *In Retrospect: The Tragedy and Lessons of Vietnam* (New York: Times Books, 1995), 11.

31. David Edgerton, "De l'innovation aux usages: Dix thèses éclectiques sur l'histoire des techniques," *Annales. Histoire, Sciences Sociales* 53, 4–5 (1998): 815–38.

32. Arnold Thackray, Jeffrey L. Sturchio, P. Thomas Carroll, and Robert Bud, *Chemistry in America, 1876–1976: Historical Indicators* (Dordrecht: Reidel, 1985).

33. Glenn C. Williams and Edward Vivian, "Pioneers in Chemical Engineering at MIT," in Furter, ed., *History of Chemical Engineering*, 113–19; MIT, "Reports of the President, 1888–1889."

34. MIT, "Report of the President for the Academic Year 1887–1888." Cited by Van Antwerpen in Furter, ed., *History of Chemical Engineering*, 6.

35. The catalogues of these universities are very clear on this point. See Trescott, "Unit Operations in the Chemical Industry: An American Invention in Modern Chemical Engineering," in Furter, ed., *Century of Chemical Engineering*, 6.

36. *Catalogue of the University of Pennsylvania for 1892–1893*, Edgar Fahs Smith Collection, University of Pennsylvania Library.

37. See Samuel Haber, *Efficiency and Uplift: Scientific Management in the Progressive Era, 1890–1920* (Chicago: University of Chicago Press, 1964); Daniel Nelson, *Frederick W. Taylor and the Rise of Scientific Management* (Madison: University of Wisconsin Press, 1980); Kanigel, *One Best Way.*

38. See Hugh G. H. Aitken, *Scientific Management in Action: Taylorism at Watertown Arsenal, 1908–1915* (Princeton: Princeton University Press, 1985).

39. See Layton, *Revolt of the Engineers.*

40. DuPont's personnel registers for this period did not indicate a person's special field of engineering, which is why it is difficult to know exactly how many chemical engineers it employed before 1920.

41. John Charles Rumm, "Mutual Interest: Managers and Workers at the Du Pont Company, 1802–1915" (Ph.D. diss., University of Delaware, 1989).

42. Charles Reese, "Twenty-five Years' Progress in Explosives," *Journal of the Franklin Institute* 198 (124), HML.

43. Rumm, "Mutual Interest," 21.

44. Donald Stabile, "The Du Pont Experiments in Scientific Management: Efficiency and Safety, 1911–1919," *Business History Review* 61 (1987): 365–86.

45. Chandler, *Visible Hand*, 438. Also Michael Massouh, "Technological and Managerial Innovation: The Johnson Company, 1883–1898," *Business History Review* 50, 1 (1976): 46–68.

46. HML, ser. 2, Acc. 518, box 1005. Letter also cited by Stabile, "Du Pont Experiments."

47. Stabile, "Du Pont Experiments," 372.

48. Ibid., 378.

49. Alfred D. Chandler Jr. and Stephen Salsbury, *Pierre S. du Pont and the Making of the Modern Corporation* (New York: Harper & Row, 1971), 360–63.

50. B. P. Foster, "History of the Engineering Department of E. I. Du Pont de Nemours & Company," HML, Pictorial Division; J. Thompson Brown, "Retirement Speech of [Everett] Ackart," 30 August 1946, HML, Acc. 1005.

51. Engineering Department personal files; Alfred D. Chandler Jr., interview of Everett Ackart, HML, Acc. 1689, 7.

52. Robert Burns MacMullin, *Odyssey of a Chemical Engineer: The Autobiography of Robert Burns MacMullin* (Lewiston, N.Y.: author, 1983).

53. Ibid., 74.

54. Aniline was used in the manufacturing of indigo and also diphenylamine, the stabilizer for smokeless powder.

55. MacMullin, *Odyssey of a Chemical Engineer*, 81. MacMullin eventually fought in Europe as a member of the 30th Engineer Regiment (Gas and Flame), which in mid-1918 was redesignated the 1st Gas Regiment.

56. See Jo Anne Yates, *Control Through Communication: The Rise of Systems in American Management* (Baltimore: Johns Hopkins University Press, 1989).

57. Zunz, *Making America Corporate*, 76.

58. Chandler, *Strategy and Structure;* Ernest Dale and Charles Meloy, "Hamilton M. Barksdale and the Du Pont Contribution to Systematic Management," *Business History Review* 34 (Summer 1962); Noble, *America by Design*, 282.

59. HML, Acc. 1662, box 8.

60. See Alfred Sloan, *Adventures of a White-Collar Man* (New York: Doubleday, 1941) and *My Years with General Motors* (New York: Doubleday, 1964).

61. Chandler, *Strategy and Structure*, 73.

62. Terry S. Reynolds, *75 Years of Progress: A History of the American Institute of Chemical Engineers* (New York: AIChE, 1983), 1–25.

63. Charles F. McKenna, "The Justification of the American Institute of Chemical Engineers," *AIChE Transactions* 1 (1908): 2.

64. Terry S. Reynolds, "Defining Professional Boundaries: Chemical Engineering in the Early Twentieth Century," in Reynolds, ed., *Engineer in America*, 354.

65. Reynolds, *75 Years of Progress*, 8.

66. A complete list can be found in Hounshell and Smith, *Science and Corporate Strategy*, appendix III, 608–9.

67. Fin Sparre to C. M. Barton, 30 August 1910, HML, ser. 2, Acc. 2, box 253.

68. Charles Stine, "Chemical Engineering in Modern Chemical Industry" (Niagara Falls, 24 August 1928), HML, Acc. 1706.

69. F. B. Holmes, memorandum to W. M. Whitten, 7 May 1920, HML, Acc. 1784.

70. On the reorganization of 1920, see Chandler, *Strategy and Structure*, 96–110.

71. Henry Mintzberg, *The Structuring of Organizations: A Synthesis of the Research* (Englewood Cliffs, N.J.: Prentice-Hall, 1979); Patrick Fridenson, "Les organisations: Un nouvel objet," *Annales. Histoire, Sciences Sociales* 44, 6 (November–December 1989): 1461–77.

72. Luc Boltanski, *Les cadres: La formation d'un groupe social* (Paris: Minuit, 1982).

73. MacMullin, *Odyssey of a Chemical Engineer*, 72ff.

74. *AIChE Bulletin* 24 (1921): 53.

75. U.S. Department of Commerce, Bureau of the Census, *Statistical Tables, 1900–1970*, ser. D 257, 140.

76. See Robert J. Thomas, *What Machines Can't Do: Politics and Technology in the Industrial Enterprise* (Berkeley: University of California Press, 1995).

77. Arthur D. Little, "Report of the AIChE Committee on Chemical Engineering Education," HML, Longwood.

78. Reynolds, *75 Years of Progress*, 12–14.

79. *Transactions in Chemical Engineering*, the official publication of the AIChE, for instance, underwent a definite change after 1920.

80. Little, "Report of the AIChE Committee" (cited n. 77 above), 8.

81. MacMullin refers to this incident in *Odyssey of a Chemical Engineer*, 76.

82. This process replaced the Leblanc and Solvay processes. See Aftalion, *Histoire de la chimie*, 62–65; Reese, "Does Chemical Engineering Pay?" HML, Acc. 1706, 5.

83. On batch processing, see Philip Scranton, *Proprietary Capitalism: The Textile Manufacture at Philadelphia, 1800–1885* (Cambridge: Cambridge University Press, 1983). As American chemical engineers who visited French chemical plants in the 1920s pointed out, the main differentiating factor between American and French plants was that the Americans aimed to achieve continuous-flow production. Alfred D. Chandler Jr., interview of Frank Sanderson McGregor, HML, Acc. 1662.

84. William H. Walker, Warren K. Lewis, and William H. McAdams, *Principles of Chemical Engineering* (New York: McGraw-Hill, 1921).

85. See David A. Hounshell, *From the American System to Mass Production: The Development of Manufacturing Technology in the United States* (Baltimore: Johns Hopkins University Press, 1984).

86. Henry Ford, "Mass Production," *Encyclopedia Britannica*, 13th ed. (1926), 2: 821–23, cited by Hounshell, *From the American System to Mass Production*, 217. The article was written by William Cameron.

87. Edward Filene, *The Way Out: A Forecast of Coming Changes in American Business and Industry* (Garden City, N.Y.: Doubleday, 1925). It should be kept in mind that mass production does not necessarily mean standardization of products, for the consumers wanted both massive and varied supplies of goods, a fact of which the strategists at DuPont and General Motors were very well aware.

88. Cecilia Tichi, *Shifting Gears: Technology, Literature, Culture in Modernist America* (Chapel Hill: University of North Carolina Press, 1987).

89. Siegfried Giedion, *Mechanization Takes Command: A Contribution to Anonymous History* (1948; repr., New York: Oxford University Press, 1969). According to Giedion, the assembly line was the culmination of the American manufacturing system. Nowadays, historians of technology are more likely to insist on the essentially unique nature of the assembly lines at the Ford plants and on the profound impact Fordism had in the twentieth century. Note also that the term "Fordism" as I am using it here does not refer to an explicit doctrine but rather to a set of practices improvised and improved upon by Ford and his men, as Charles Sorenson explains in his memoirs, *My Forty Years with Ford* (New York: Norton, 1956).

90. Jean-Claude Guédon, "Conceptual and Institutional Obstacles to the Emergence of Unit Operations in Europe," in Furter, ed., *History of Chemical Engineering*, 49.

91. On Perkin, see Anthony S. Travis, *The Rainbow Makers: The Origins of the Synthetic Dyestuffs Industry in Western Europe* (Bethlehem, Pa.: Lehigh University Press, 1993).

92. Aftalion, *Histoire de la chimie*, 108.

93. See D. C. Freshwater, "George E. Davis, Norman Swindlin, and the Empirical Tradition in Chemical Engineering," in Furter, ed., *History of Chemical Engineering*, 97–111.

94. J. Cathala, "Chemical Engineering in France," in Edgar L. Piret, *Chemical Engineering Around the World* (New York: AIChE, 1958), 139–63.

95. Mary Jo Nye, *Science in the Provinces: Scientific Communities and Provincial Leadership in France, 1860–1930* (Berkeley: University of California Press, 1986); André Grelon, ed., *Les ingénieurs de la crise: Titre et profession entre les deux guerres* (Paris: École des hautes études en sciences sociales, 1986).

96. Paul Pascal, *Synthèses et catalyses industrielles* (Paris: Hermann, 1925).

97. Guédon, "Conceptual and Institutional Obstacles," 56–62, and André Thépot, "The Evolution of Chemical Engineering in the Heavy Inorganic Chemical Industry in France during the Nineteenth Century," in Furter, ed., *Century of Chemical Engineering*, 55–64.

98. On these productivity commissions, see Anthony Rowley, "Taylorisme et missions de productivité aux États-Unis au lendemain de la Seconde Guerre Mondiale," in Maurice de Montmollin and Olivier Pastré, eds., *Le Taylorisme* (Paris: La Découverte, 1984); also Vincent Guigueno, "L'éclipse de l'atelier: Les missions françaises de productivité aux États-Unis dans les années 1950" (Diplôme d'études approfondies thesis, Université de Marne-la-Vallée, Cité Descartes, Paris, 1994).

99. Karl Schoenemann, "The Separate Development of Chemical Engineering in Germany," in Furter, ed., *History of Chemical Engineering*, 249–72.

100. John J. Beer, *The Emergence of the German Dye Industry* (Urbana: University of Illinois Press, 1959).

101. Victor Cambon, *Les derniers progrès de l'Allemagne* (Paris: Pierre Roger, 1914).

102. Karl Schoenemann, "Chemical Engineering in Germany," in Piret, ed., *Chemical Engineering Around the World*, 198.

103. Schoenemann, "Separate Development," in Furter, ed., *History of Chemical Engineering*, 249–72.

104. Haynes, *American Chemical Industry*, 1: 265–66.

105. Chandler interview of Everett Ackart, HML, Acc. 1689, 11.

106. Reynolds, *75 Years of Progress*, 13–15.

107. Charles Reese, "How Can the Industries Aid Engineering Education?" (address to the American Society of Mechanical Engineers, University of Virginia, 5 May 1922), HML, Acc. 1706.

108. On the poison gases, see Ludwig F. Haber, *The Poisonous Cloud: Chemical Warfare in the First World War* (Oxford: Clarendon Press, 1986), and Olivier Lepick, *La Grande Guerre chimique, 1914–1918* (Paris: Presses universitaires de France, 1998). Ludwig Haber is the son of Fritz Haber, the German chemist who synthesized ammonia and developed the first military poison gases. Haber only briefly refers to the manufacture of military poison gas in the United States. See also therefore Daniel P. Jones, "The Role of Chemists in Research on War Gases in the United States During World War I" (Ph.D. diss., University of Wisconsin, 1969) and the same author's "Chemical Warfare Research During World War I," in John Paranscandola and James C. Whorton, eds., *Chemistry and Modern Society: Historical Essays in Honor of Aaron J. Ihde* (Washington, D.C.: American Chemical Society, 1983). The discussion of this topic here is based on the work of Haber and Jones and on the archives of the Chemical Warfare Service, National Archives, College Park, Md., RG 156 and 175.

109. There are many literary works evoking the use of poison gas during World War I, such as Henri Barbusse's *Le feu* (1916; trans. as *Under Fire*), and *Les champs d'honneur* by Jean Rouault (Paris: Minuit, 1992), with its moving portrayal of gassed soldiers and their agonies. Nonetheless, it should be noted that gas did not play a very important military role during the war, and that it was, as Haber puts it, "a military failure." The use of poison gas in warfare was outlawed by the Geneva gas protocol in 1925.

110. James Hershberg, *James B. Conant: Harvard to Hiroshima and the Making of the Nuclear Age* (New York: Knopf, 1993), 35–48.

111. D. P. Jones, "Role of Chemists," 98–103. Chemists could avoid conscription by contacting the American Chemical Society, which then transferred them to the poison gas program.

112. See National Archives, College Park, Md., RG 175, entry 11, for a report on the building of the arsenal.

113. Ibid., Technical Files, ser. 5, 6, 6A.

114. Dana J. Demorest, "The Chemical Engineering of Poison Gas Manufacture," *Chemical Warfare* 4 (1919): 3.

115. National Archives, College Park, Md., Construction Report, RG 175, entry 11, 45.

116. D. P. Jones, "Role of Chemists," 133.

117. D. P. Jones, "Chemical Warfare Research During World War I," 177.

118. Ibid., 189.

119. *Chemical Warfare* 13 (January 1927), and "Chemical Warfare Against a Semi-Civilized Enemy With Particular Reference to Its Use Against the Moro," *Chemical Warfare,* January 1930.

120. Amos Fries, "Chemical Warfare Inspires Peace," *Chemical Warfare* 6 (1921): 3.

121. M. E. Barker, "The Chemical Warfare School of Applied Chemistry at the Massachusetts Institute of Technology," *Chemical Warfare,* July 1928, 372–73, and National Archives, College Park, Md., RG 175, entry 8.

122. John Woodhouse, interview 2, 32, HML. Interviews are in HML Acc. 1878 unless otherwise indicated.

123. Robert Pigford, interview, HML, 9.

124. "Look Out for Gas!" *Time* 15 (June 1942): 23. I thank Ed Russell for having brought this quotation to my attention.

125. John H. Hammond, "The Chemical Engineer," in Dugald C. Jackson and W. Paul Jones, eds., *The Profession of Engineering* (New York: John Wiley & Sons, 1929), 107.

126. Christophe Lécuyer, "MIT, Progressive Reform and Industrial Service," *Historical Studies in Physical Sciences* 26, 1 (1995): 35–38; John Servos, *Physical Chemistry from Ostwald to Pauling: The Making of a Science in America* (Princeton, N.J.: Princeton University Press, 1990); "W. Bernard Carlson, Dugald C. Jackson and the MIT GE Cooperative Course," in Reynolds, ed., *Engineer in America,* 367–98.

127. Noble, *America by Design,* 167–223.

128. Figures taken from *Chemical Engineering Progress* 2 (1969).

129. The data specifically relating to MIT are taken from articles and books mentioned in the following notes and from the annual reports of the president of MIT, an extremely valuable source of information. I thank Christophe Lécuyer for his advice in this matter.

130. On the early years of MIT, see Christophe Lécuyer, "MIT, Progressive Reform, and 'Industrial Service,' 1890–1920," *Historical Studies in the Physical Sciences* 26, 1 (1995): 35–88; Geiger, *To Advance Knowledge;* and Samuel Prescott, *When MIT Was Boston Tech, 1861–1916* (Cambridge, Mass.: Technology Press, 1954).

131. Lécuyer, "MIT, Progressive Reform, and 'Industrial Service.'"

132. Servos, *Physical Chemistry from Ostwald to Pauling,* 110ff.

133. On Morris Cooke, see Layton, *Revolt of the Engineers.* According to Cooke, the engineers had to choose between serving their employers or serving the public. Having reorganized the Public Works Department of Philadelphia, Cooke led the Association of Mechanical Engineers (ASME) in the Progressive "Revolt of the Engineers," a movement opposed to big business.

134. Brochure of the Department of Chemical Engineering, 1920, cited in John Mattill, *The Flagship: The MIT School of Chemical Engineering Practice, 1916–1991* (Cambridge, Mass.: David H. Koch School of Chemical Engineering Practice, MIT, 1991).

135. See Robert Kargon, *The Rise of Robert Millikan: Portrait of a Life in American Science* (Ithaca, N.Y.: Cornell University Press, 1982).

136. MIT, "Annual Reports of the President," 1910–11 and 1911–12.

137. Alfred D. Chandler Jr., interview of Frank Sanderson McGregor, HML, Acc. 1689, 37.

138. Noble, *America by Design,* 142–44, and William A. Walker, "The Division of

Industrial Cooperation and Research at MIT," *Journal of Industrial and Engineering Chemistry* 12 (April 1920): 394.

139. In 1932, the Technology Plan became the Division of Industrial Cooperation, chaired by Karl Compton. After the war it was renamed the Division of Sponsored Research and expanded its relations with the government and the military. As Christophe Lécuyer has shown, the choice of Karl Compton as its new president in 1929 testified to MIT's awareness of a decline that threatened its reputation, and to its decision to refocus on scientific programs better suited to its ambitions. Yet it appears that from about 1915 to the end of the 1920s, MIT, faced with heavy expenses and crippled with debts, was tempted to place itself entirely and wholeheartedly at the service of industry, in this instance following the path opened by the forceful personality of William Walker.

140. Mattill, *Flagship*, and MIT, "Annual Report of the President," 1916–17.

141. W. Bernard Carlson, "Dugald C. Jackson and the MIT-GE Cooperative Course," in Reynolds, ed., *Engineer in America,* 367–98.

142. "Report of the Committee on Chemical Engineering Education of the American Institute of Chemical Engineers," 1922, HML, Longwood.

143. E. K. Bolton, "Development of Research in the Du Pont Company," 20 December 1932, HML, ser. 2, 2, box 1035, and Cole Coolidge, "Organization of Du Pont Research, 23 June 1952," Acc. 1850.

144. MIT Technology Loan Fund, 18 January 1932, HML, Longwood, 367–68.

145. Report of the Visiting Committee of the Department of Chemistry and Chemical Engineering, 26 March 1923, HML, Longwood, 367.

146. "Semi-annual Report of the Chemical Department, January–June 1920," 28 September 1920, 69, HML, Acc. 1784.

147. MIT, President's Report, Meeting of the Corporation, 16 October 1925, 19, 22, 23, HML, Longwood, 367.

148. MIT, Meeting of the Corporation, 8 January 1930, HML, Longwood, 367.

149. HML, Acc. 1813, box 2.

150. Reese, "How Can the Industries Aid Engineering Education?" 3.

151. Crawford Greenewalt, letter to Roger Williams, 27 April 1948, HML, Acc. 1814, box 30.

152. Robert Clothier of Rutgers University, letter to Lammot du Pont, and memos between du Pont and Stine, HML, Acc. 1622, box 17.

153. F. C. Evans, "Scouting for Chemists in our Colleges," HML, Acc. 1784, box 19.

154. A good discussion of the activities of Adams and Marvel is to be found in Hounshell and Smith, *Science and Corporate Strategy,* 297–98. See also "Carl Shipp Marvel, Polymer Pioneer" (biographical sketch), in HML, Acc. 1784, box 18, and a biographical sketch of Roger Adams in W. D. Miles, ed., *American Chemists and Chemical Engineers* (Washington, D.C.: American Chemical Society, 1976), 4–5.

155. John Woodhouse, interview 1, 18.

156. Chaplin Tyler, interview 1, 8.

157. Reynolds, "Chemical Engineering in the Early Twentieth Century," in Reynolds, ed., *Engineer in America,* 364.

T W O :   From Ammonia to Nylon

1. Chandler, *Strategy and Structure*, 78ff.

2. Ibid., 112–13. See also Olivier Zunz, *L'Amérique en col blanc: L'invention du tertiaire, 1870–1920* (Paris: Belin, 1992), 108–24.

3. Hounshell and Smith, *Science and Corporate Strategy*, 109.

4. Alfred D. Chandler Jr., interview of Elmer Bolton, HML, Acc. 1689, 13.

5. This is the perspective of Charles Cheape in *Strictly Business: Walter Carpenter at Du Pont and General Motors* (Baltimore: Johns Hopkins University Press, 1995); see also Fridenson, "Organisations," 1474.

6. Hounshell and Smith, *Science and Corporate Strategy*, 149–50.

7. Kathryn Steen, "Wartime Catalyst and Postwar Reaction: The Making of the United States Synthetic Organic Chemicals Industry, 1910–1930" (Ph.D. diss., University of Delaware, 1995), 391.

8. Aftalion, *Histoire de la chimie*, 76–78; also Brooks Darlington, "The Story of a Yarn," *Du Pont Magazine,* August 1934, 10–15.

9. Ferdinand Schulze, *The Technical Division of the Rayon Department, 1920–1951* (Wilmington, Del.: E. I. Du Pont de Nemours, 1952); Hounshell and Smith, *Science and Corporate Strategy,* 166; Janice Reed, "The Progress of Rayon," *Commerce Monthly* (National Bank of Commerce) 10 (4 August 1928).

10. See Samuel Hollander, *The Sources of Increased Efficiency: A Study of the Du Pont Rayon Plants* (Cambridge, Mass.: MIT Press, 1965).

11. Cellophane was invented by a Swiss chemist, Jacques Brandenberger, who worked for a small firm, the Blanchisserie et Teinturerie of Thaon. This company was later acquired by the Comptoir des textiles artificiels, which founded a subsidiary called La Cellophane. See *Cellophane 25th Anniversary, 1924–1949* (DuPont pamphlet, 1949), HML; also Aftalion, *Histoire de la chimie*, 78–79.

12. Hughes, *American Genesis*, 175; Hounshell and Smith, *Science and Corporate Strategy.*

13. MacMullin, *Odyssey of a Chemical Engineer;* Chandler interview of Elmer Bolton; also Bolton, "Research Efficiency," 16 July 1920, HML, Acc. 1784.

14. Chandler interview of Elmer Bolton, 16.

15. Recall that nitrogen cannot be directly assimilated by plants (except leguminous plants). Plants obtain nitrogen in a different form, as nitrates.

16. These reserves are found in Chile's Atacama desert; sodium nitrate is water soluble, which is why it is found in large quantities only in desert zones. These deposits of Chile saltpeter became particularly attractive from the late 1870s on, when an industrial process for treating them had been developed and when Chile had gained definitive control of this region following the "Pacific War" against Bolivia (1879–82). DuPont and the Guggenheim family controlled a number of saltpeter mines.

17. Harold Tongue, *The Design and Construction of High Pressure Chemical Plants* (London: Chapman & Hall, 1934).

18. High pressure favored the reaction. The catalyst is used to break the very strong connection between the two nitrogen (N) atoms, which are attached to each other by a triple bond.

19. Max Appl, "The Haber-Bosch Process and the Development of Chemical Engineering," in Furter, ed., *Century of Chemical Engineering*, 29–54; Kenzi Tamaru, "The

History of the Development of Ammonia Synthesis," in J. R. Jennings, ed., *Catalytic Ammonia Synthesis: Fundamentals and Practice* (New York: Plenum Press, 1991), 1–18. Note that the Haber process is used to this day, although it has been improved by building better catalyzers and lowering the pressures, which also lowered production costs. On Carl Bosch, see Peter Hayes, *Industry and Ideology: I.G. Farben in the Nazi Era* (New York: Cambridge University Press, 1987). On Haber, see Dietrich Stolzenberg, *Fritz Haber: Chemiker, Nobelpreisträger, deutscher Jude* (Weinheim: VCH, 1994); also Margit Scöllösi-Janze, *Fritz Haber, 1868–1934: Eine Biographie* (Munich: H. C. Beck, 1998).

20. Victor Cambon, *L'Allemagne nouvelle* (Paris: Pierre Roger, 1923), 76; Louis Hackspill, *L'azote: La fixation de l'azote atmosphérique et son avenir industriel* (Paris: Masson, 1922), and Paul Pascal, *Synthèses et catalyses industrielles* (Paris: Hermann, 1925).

21. Cambon, *Allemagne nouvelle*, 155.

22. "Billingham" Report, HML, ser. 2, box 1036.

23. On DuPont's activities in Chile, see Cheape, *Strictly Business*.

24. Ludwig Haber, *Chemical Industry*, 205–6.

25. Edward P. Bartlett, "The Chemical Division at the Du Pont Ammonia Department, 1924–1935," HML, ser. 2, box 91.

26. "We Hoped to Make Money," *Fortune*, August 1932, 58ff. Sparre made a dubious strategic choice when he preferred the Claude process to the Haber process—but this did not affect his career, for he remained in charge of the Development Department until his retirement in 1944.

27. Ibid.

28. In 1928, the French firm Kuhlmann hired Nitrogen Engineering Corporation to built its synthetic ammonia plant near Paris.

29. Haynes, *American Chemical Industry*, 88; also letter from Lammot du Pont, 20 August 1931, HML, Acc. 1813, box 7.

30. F. C. Zeisberg, "The Haber Process of Synthesizing Ammonia," 27 July 1926, HML, Acc. 1784.

31. Bartlett, "Chemical Division." On the war profits, see Chandler and Salsbury, *Pierre S. du Pont and the Making of the Modern Corporation*, 428.

32. Ludwig Haber, *Chemical Industry*, 91.

33. They are found in HML, Acc. 1813.

34. Walter Carpenter, letter to Frederic A. Wardenburg, 12 October 1939, HML, ser. 2, box 1037.

35. The information about Roger Williams given here is from HML, History TNX, Acc. 1957, ser. 1, A, box 1, p. 6; Leslie R. Groves, *Now It Can Be Told: The Story of the Manhattan Project* (New York: Harper & Row, 1962), 52; William Haynes, ed., *Chemical Who's Who: Biography in Dictionary Form of the Leaders in Chemical Industry, Research, and Education*, 3d ed. (New York: Lewis Historical Publishing Co., 1951); and interview of Williams by Chandler, HML, Acc. 1689.

36. Roger Williams to Hamilton Bradshaw, "Recent Developments Concerning the Haber Process," 21 November 1919, HML, Acc. 1784.

37. Hounshell and Smith, *Science and Corporate Strategy*, 187.

38. Haynes, ed., *Chemical Who's Who*; also "Ammonia Department, Organization," 31 December 1935, HML, Acc. 1784, box 23.

39. Haynes, ed., *Chemical Who's Who*.

40. J. K. Smith, interview of Chaplin Taylor, 11.

41. John Woodhouse, interview 3, 11.

42. John Woodhouse, interview 1, 9.

43. HML, 1813, box 3.

44. As early as 1925, BASF had begun producing synthetic methanol at its Leuna plant, which was later to make synthetic gasoline for the Wehrmacht and was leveled by Allied bombing in 1945. As for DuPont, its methanol production increased from 2,000 tons in 1927 to 100,000 tons in 1936.

45. More generally speaking, DuPont's total sales had gone from $214 million in 1929 to $127 million in 1932. "Permanent Investments—Eleven Largest Plants—31 August 1937," HML, Acc. 1813, box 26.

46. John Woodhouse, interview 1, 21.

47. John Woodhouse, interview 2, 20.

48. Chaplin Tyler, interview 1, 19.

49. "Memorandum for Group Leaders," HML, 1784, box 19.

50. Crawford Greenewalt, "Research, the Great Reformer," 10 May 1951, HML, Acc. 1410, box 61.

51. Haynes, ed., *Chemical Who's Who*; Donald F. Boucher, *History of the Engineering Technology Laboratory, Part I, The First 25 Years, 1929–1953* (Wilmington, Del.: Du Pont Engineering Department, 1980).

52. These biographical details are from the interviews Greenewalt gave Hounshell and Smith in 1982–83, hereafter cited as Greenewalt, interviews 1, 2, 3, and 4 (HML, Acc. 1878); also from *Time*, 16 April 1951.

53. Greenewalt, interview 1, 4.

54. Greenewalt, interview 1, 2.

55. *Time*, 16 April 1951.

56. Greenewalt, interview 1, 8.

57. Greenewalt, interview 1, 13, 16, 20.

58. John Woodhouse, interview 1, 27.

59. Greenewalt, interview 1, 15.

60. Henry Mintzberg, *Power in and Around Organizations* (Englewood Cliffs, N.J.: Prentice-Hall, 1983), esp. pt. 2, "The Internal Coalition." See also R. M. Canter, *Men and Women of the Corporation* (New York: Basic Books, 1977).

61. Mintzberg, *Power*, 171ff., "The Systems of Politics."

62. Michel Winok, "Les générations intellectuelles," *Vingtième Siècle* 22 (April–June 1989): 43–60.

63. On the mathematical modeling of the unit operations, see R. Byron Bird, Warren E. Steward, and Edwin L. Lightfoot, "The Role of Transport Phenomena in Chemical Engineering Teaching and Research: Past, Present, and Future," in Furter, ed. *History of Chemical Engineering*, 153–65.

64. Jean-Claude Guédon, "From Unit Operations to Unit Processes," in John Parascandola and James Whorton, eds., *Chemistry and Modern Society* (Washington, D.C.: American Chemical Society, 1983), 43–60.

65. Boucher, *History of the Engineering Technology Laboratory*; Allan Colburn, "A Method of Correlating Forced Convection Heat Transfer Data and a Comparison with Fluid Friction," *Transactions in Chemical Engineering* 29 (1938): 174–210.

66. Louis Theodore, *Transport Phenomena for Engineers* (Scranton, Pa.: International Textbook Co., 1971).

67. After World War II, differential equations would make it possible to use computers and cybernetic procedures in chemical engineering.

68. Greenewalt, interview 1, 15, 19.

69. For a discussion of the notion of trend, see Yves Cohen and Dominique Pestre, "Présentation," *Annales. Histoire, Sciences Sociales* 53, 4–5, *Histoire des techniques* (July–October 1998): 721–44.

70. David F. Noble, *Forces of Production: A Social History of Industrial Automation* (New York: Oxford University Press, 1984).

71. Alfred D. Chandler Jr., interview of Frank Sanderson McGregor, Acc. 1689, 4.

72. Brochures published by the Department of Public Relations, such as "Nylon, the First 25 Years," HML, Acc. 1784, box 18.

73. The photographs we have were thus not taken at the actual moment of invention and have the value of mythological representations.

74. David A. Hounshell, "Du Pont and the Management of Large-Scale Research and Development," in Peter Galison and Bruce Hevly, eds., *Big Science: The Growth of Large-Scale Research* (Stanford: Stanford University Press, 1992), 236–61.

75. Jeffrey Meikle, *American Plastic: A Cultural History* (New Brunswick, N.J.: Rutgers University Press, 1995).

76. Ibid., 289.

77. Hounshell and Smith, *Science and Corporate Strategy*, chap. 12.

78. Charles Stine to the Executive Committee, "Pure Science Work" (18 December 1926), HML, Acc. 1784, box 16.

79. Matthew Hermes, *Enough for One Lifetime: Wallace Carothers, Inventor of Nylon* (Philadelphia: Chemical Heritage Foundation and American Chemical Society, 1996); also letter from Stine to Carothers, 27 September 1927, HML, Acc. 1784, box 18.

80. Herbert Morawetz, *Polymers: The Origins and Growth of a Science* (New York: John Wiley, 1985).

81. Cf. Meikle, *American Plastic*, chap. 2.

82. Cf. Peter J. T. Norris, *Polymer Pioneers* (Philadelphia: Beckman Center for the History of Chemistry, 1990); Eduard Farber, "Leo Hendrick Baekeland," in *Dictionary of Scientific Biography* (New York: Scribner, 1970).

83. Chandler interview of Elmer Bolton, 20, HML, Acc. 1689.

84. Young Americans knew Mark well in the 1960s, since he hosted an educational show called *Continental Classroom* on NBC.

85. On synthetic rubber, see Norris, *Polymer Pioneers*.

86. E. K. Bolton, "Development of Nylon," *Industrial and Engineering Chemistry* 34 (January 1942): 53–58; HML, Acc. 2028.

87. Hounshell and Smith, *Science and Corporate Strategy*, 237–38.

88. See Leonard Reich, *The Making of American Industrial Research: Science and Business at General Electric and Bell, 1876–1926* (New York: Cambridge University Press, 1985), chap. 7.

89. Ibid., 244.

90. Bolton, "Development of Nylon," 55.

91. Madeleine Akrich, Michel Callon, and Bruno Latour, "À quoi tient le succès des innovations," *Annales des mines*, ser., *Gérer et comprendre*, June 1988, 4–17; December 1988, 14–29.

92. Ibid., June 1988, 16.

93. Hounshell and Smith, *Science and Corporate Strategy,* 308–9.

94. Ibid., 65.

95. Ibid., 183–89, and A. D. Chandler, R. D Williams, and N. B. Wilkinson, interview of Roger Williams, 20 June 1962, HML, Acc. 1689.

96. Bartlett, "Chemical Division."

97. Ibid., 56.

98. Bolton, "Development of Nylon."

99. Greenewalt, interview 1, 21.

100. Carothers and Graves, "Polyamide Fibers: Summary Report, January 1 to March 31, 1936," HML, Acc. 1784, box 16.

101. G. P. Hoff, "Visit to Experimental Station, October 28–29," 12 November 1936, HML, ser. 2, 2, box 958.

102. See Groves, *Now It Can Be Told.*

103. Chandler interview of Everett Ackart, HML, Acc. 1689.

104. "Engineering Department Study, Report D–1240, 17 November 1947," HML, Acc. 1801, box 28.

105. Bolton, "Development of Nylon," 56, HML, Acc. 2028.

106. Ibid., 51.

107. On the "nylon model" see Hounshell, "Du Pont and the Management of Large-Scale Research and Development." On the star quality of chemistry, see esp. the DuPont archival material on the DuPont stand at the 1939 Golden Gate International Exposition in San Francisco, HML, Acc. 1410, box 44.

108. "Du Pont Announces a New Synthetic Rubber at Akron Group Meeting," *Rubber Age* 12 (November 1931), in HML, Acc. 1850.

109. Archives of DuPont's Public Relations Department, Acc. 1410, boxes 49 and 50.

110. Charles Stine, "Training Tomorrow's Industrial Leaders" (annual meeting of the AIChE, Philadelphia, 11 November 1938), HML, Acc. 1706, 12–13.

111. Ibid., 15.

112. Introduction to the Chandler Conference, Charles F. Chandler Foundation, Columbia University, HML.

113. Thomas Chilton, "Engineering in the Service of Industry" (Chandler Lecture, 1939), 3–4, HML.

T H R E E : Culture and Politics at DuPont before World War II

1. Two studies have been devoted to the political activities of the du Ponts: George Wolfskill, *The Revolts of the Conservatives: A History of the American Liberty League, 1934–1940* (Boston: Houghton Mifflin, 1962), and Robert F. Burk, *The Corporate State and the Broker State: The du Ponts and American National Politics, 1925–1940* (Cambridge, Mass.: Harvard University Press, 1990).

2. Chandler and Salsbury, *Pierre S. du Pont and the Making of the Modern Corporation,* 104.

3. *Hearings Before the Subcommittee of the Committee on Appropriations, House of Representatives* (Washington, D.C.: Government Printing Office, 1907).

4. Chandler and Salsbury, *Pierre S. du Pont and the Making of the Modern Corporation,* 259ff.

5. See Thomas McCraw, *Prophets of Regulation: Charles Francis Adams, Louis D. Brandeis, James M. Landis, Alfred E. Kahn* (Cambridge, Mass.: Belknap Press of Harvard University Press, 1984).

6. Herbert Croly, *The Promise of American Life* (1909; repr., New Brunswick, N.J.: Transaction, 1993); David Potter, *People of Plenty: Economic Abundance and the American Character* (Chicago: University of Chicago Press, 1954).

7. These federal commissions have been seen in two different perspectives. One is that of Martin Sklar in *The Corporate Reconstruction of American Capitalism, 1890–1916: The Market, the Law, Politics* (New York: Cambridge University Press, 1988); the other is that of Peter Temin in *Taking Your Medicine: Drug Regulation in the United States* (Cambridge, Mass.: Harvard University Press, 1980); see also Louis Galambos and Joseph Pratt, *The Rise of the Corporate Commonwealth: United States Business and Public Policy in the Twentieth Century* (New York: Basic Books, 1988), chap. 3.

8. The text of the Sherman Act is reproduced in many books, e.g., Galambos and Pratt, *Rise of the Corporate Commonwealth*, 60.

9. Dutton, *Du Pont: One Hundred and Forty Years*, 119ff.

10. See Naomi Lamoreaux, *The Great Merger Movement in American Business, 1895–1904* (New York: Cambridge University Press, 1985), and Sklar, *Corporate Reconstruction*.

11. The anti-trust legislation was not directed against monopolies as such but only against those that had been formed by illegal methods. See Sklar, *Corporate Reconstruction*, 179ff.

12. David Graham Philips, *The Treason of the Senate* (1953; repr., Chicago: Quadrangle Books, 1964).

13. Democrat Josiah O. Wolcott, elected senator in 1916, was named chancellor of the Delaware Court of Chancery by the Republican governor, William D. Denny, who then appointed Republican T. Coleman du Pont to take over Wolcott's Senate seat. Newspapers called this "Delaware's Dirty Deal."

14. Robert Wiebe, "The House of Morgan and the Executive, 1905–1913," *American Historical Review* 65 (October 1959): 49–60, and id., *The Search for Order, 1877–1920* (New York: Hill & Wang, 1967).

15. Chandler and Salsbury, *Pierre S. du Pont and the Making of the Modern Corporation*, 248.

16. Figures compiled by Chandler and Salsbury, ibid., appendix.

17. The details of the contract can be found in the report of the Nye Committee, "Construction of Explosives Plant near Nashville, Tennessee" (Nye Committee Investigation, pt. 14, 14 December 1934).

18. U.S. Congress, "House Report 998," 66th Cong., 2d sess.

19. Chandler and Salsbury, *Pierre S. du Pont and the Making of the Modern Corporation*, 396.

20. Gerard Colby, *Du Pont Dynasty* (Secaucus, N.J.: L. Stuart, 1984), and see also id., *Du Pont: Behind the Nylon Curtain* (Englewood Cliffs, N.J.: Prentice-Hall, 1974); Graham D. Taylor and Patricia E. Sudnick, *Du Pont and the International Chemical Industry* (Boston: Twayne, 1984).

21. *New York World*, 18 October 1916, cited in Chandler and Salsbury, *Pierre S. du Pont and the Making of the Modern Corporation*, 398.

22. The Progressive reforms are conveniently summarized in Alan Brinkley, *The*

Unfinished Journey: A Concise History of the American People (New York: Knopf, 1993). See also Pierre-Yves Nouailhat, *Les États-Unis, l'avènement d'une puissance mondiale, 1896–1933* (Paris: Éditions Richelieu, 1973).

23. On Prohibition in general, see John Kobler, *Ardent Spirits: The Rise and Fall of Prohibition* (1973; repr. New York: Da Capo Press, 1993). See also Jean-Pierre Martin, *La vertu par la loi: La prohibition aux États-Unis, 1920–1933* (Dijon: Université de Bourgogne, 1992).

24. This association was never as popular as the Anti-Saloon League, the rival *dry* organization, which at the time received most of its funds from members of modest means, although Henry Ford and John D. Rockefeller belonged to it as well. Many businessmen supported the AAPA, among them the presidents of U.S. Steel, Westinghouse, and the Pennsylvania Railroad, and the heads of many important banks.

25. Pierre S. du Pont, *A Plan for Distribution and Control of Intoxicating Liquors in the United States, Suggested by Pierre S. du Pont* (n.p., September 1930), HML, Longwood, PSDP, ser. A, file 1023.

26. Thomas Ferguson, "Industrial Conflict and the Coming of the New Deal: The Triumph of Multinational Liberalism in America," in Steve Frazer and Gary Gerstle, eds., *The Rise and Fall of the New Deal Order, 1930–1980* (Princeton, N.J.: Princeton University Press, 1989), 3–31.

27. See, e.g., the correspondence between Pierre du Pont and Jouett Shouse, 9 September and 16 October 1931, HML, Longwood, PSDP, ser. A, file 1023. Before 1933, the Democratic Party was deeply divided among different factions, such as the conservative Southern Democrats, Progressives, Democrats with close ties to business, etc.

28. Pierre du Pont to Matthew Wohl, 7 November 1932, HML, Longwood, ser. A, file 1023.

29. Alan Brinkley, *The End of Reform: New Deal Liberalism in Recession and War* (1995; repr. New York: Vintage Books, 1996), 25.

30. See Thomas McCraw, "The New Deal and the Mixed Economy," in Harvard Sitkoff, ed., *Fifty Years Later: The New Deal Evaluated* (New York: Knopf, 1985), and Bernard Bellush, *The Failure of the NRA* (New York: Norton, 1975). Correspondence between Pierre du Pont and Roosevelt, 13 and 20 November 1933, HML, Longwood, ser. A, file 765. Swope's proposal is formulated in his book *The Swope Plan* (New York: Business Bourse, 1931). This book represents the views of a great many leaders of big industry who favored a "rationalization" of the market and the setting up of industrial cartels that would guarantee stable prices and some social benefits for the workers concerned.

31. Thomas Ferguson, "From Normalcy to New Deal: Industrial Structure, Party Competition, and American Public Policy in the Great Depression," *International Organization* 38, 1 (1984): 41–94. The Business Advisory Council remained fairly marginal, and an economy regulated by big business was never more than an illusion shared by few New Dealers. See Brinkley, *End of Reform,* 42–47.

32. Brinkley, *End of Reform,* 4.

33. William Stayton, "Memorandum Concerning the Activities of the American Liberty League from Its Organization (August 1934) to June 1938," 5, HML, Acc. 771, and "Certificate of Formation of American Liberty League," Acc. 771, box 1. Stayton burned the archives of the Liberty League in 1936 after he had tried unsuccessfully to donate them to the Library of Congress. Hence, all that is available are the brochures of

the League and the personal correspondence of certain of its members—among them the du Ponts.

34. Burk, *Corporate State,* 143.

35. *Newsweek,* 26 August 1934, 7.

36. An example is James Farley, chairman of the Democratic Party, speaking at the Democratic Convention at Philadelphia in June 1936.

37. John E. Wilz, *In Search of Peace: The Senate Munition Inquiry* (Baton Rouge: Louisiana State University Press, 1963). And see Report of the Special Committee on Investigation of the Munitions Industry (The Nye Report), U.S. Congress, Senate, 74th Cong., 2d sess., February 24, 1936, 3–13.

38. H. C. Englebrecht and F. C. Hanighen, *Merchants of Death: A Study of the International Armament Industry* (New York: Dodd, Mead, 1934), and Stewart D. Brandes, *Warhogs: A History of War Profits in America* (Lexington: University Press of Kentucky, 1997).

39. *Business Week,* 15 September 1934, 20.

40. In Delaware, for instance, huge advertising campaigns were launched by Brooks Darlington, a former journalist who had become the League's public relations director. See Darlington to Pierre du Pont, 2 October 1934, HML, Acc. 771, box 1.

41. Wolfskill, *Revolt of the Conservatives,* 248.

42. Burk, *Corporate State,* 252, 279.

43. Charles Michelson, *The Ghost Talks* (New York: G. P. Putnam's Sons, 1944), 144.

44. Wolfskill, *Revolt of the Conservatives,* 177.

45. Serge Berstein, "Les ligues, un phénomène de droite," in Jean-François Sirinelli, ed., *Histoire des droites en France* (Paris: Gallimard, 1992), 2: 61–62.

46. On the Fédération nationale des contribuables, see Pierre Milza, "L'ultra-droite des années trente," in Michel Winock, ed., *Histoire de l'extrême-droite en France* (Paris: Seuil, 1993), 160.

47. U.S. Congress, Senate, Special Committee to Investigate Lobbying Activities, *Hearings,* 74th Cong., 2d sess., 1935–36, National Archives, RG 46, 18.67–70.

48. I am borrowing this idea from Pierre Milza.

49. On the NAM, see Robert Collins, *The Business Response to Keynes, 1929–1964* (New York: Columbia University Press, 1981), chap. 1.

50. Michelson, *Ghost Talks,* 144.

51. Meikle, *American Plastic,* 133.

52. Bruce Barton, *The Man Nobody Knows: A Discovery of the Real Jesus* (Indianapolis: Bobbs-Merrill, 1925).

53. Meikle, *American Plastic,* 135.

54. Henry B. du Pont, "Industry, Past, Present, and Future" (address, Detroit, Michigan), HML, Acc. 1410, box 61A.

55. Walter S. Carpenter, "Industry's Policy as to a Fair Distribution of Income" (13 February 1940), HML, Acc. 1410, box 61A.

56. Brinkley, *End of Reform,* 65–85.

57. Stanley Lebergott, *Pursuing Happiness: American Consumers in the Twentieth Century* (Princeton, N.J.: Princeton University Press, 1993), 61–72.

58. See Collins, *Business Response to Keynes.*

59. Raymond Clapper, "Artificial Silk Worm Developed by du Ponts," *New York World Telegram,* 17 January 1939, HML, Acc. 2028.

60. See Philip Selznick, *TVA and the Grass Roots: A Study of Politics and Organization* (1949; repr. Berkeley: University of California Press, 1980).

61. See Ferdinand Lundberg, *America's Sixty Families* (New York: Vanguard Press, 1937).

62. See Charles Beard, *An Economic Interpretation of the Constitution of the United States* (New York: Macmillan, 1935).

63. See William Appleman Williams, *The Tragedy of American Diplomacy* (New York: Dell, 1963), and Gabriel Kolko, *The Triumph of Conservatism: A Reinterpretation of American History, 1900–1916* (New York: Free Press of Glencoe, 1963).

64. See Lundberg, *America's Sixty Families.* Lundberg was a radical Progressive and a friend of Roosevelt's advisor Harold Ickes. The French expression *les deux cents familles* was used for the first time at the convention of the Parti radical in 1932.

65. See, e.g., *The Worker Magazine*, supplement to the *Daily Worker* (New York), 27 August 1950.

66. Gerard Colby, *Du Pont: Behind the Nylon Curtain* (Englewood Cliffs, N.J.: Prentice-Hall, 1974).

67. Chandler, *Visible Hand.*

68. Metropolitan Life is a good example; see Zunz, *Amérique en col blanc,* 142–59.

69. A symmetrical but inverse position is that of Robert Burk, who does take an interest in the du Ponts' political activities during the interwar period but fails to relate them in any way to the company and its employees, who are not even mentioned in Burk's *Corporate State.*

70. Thomas G. Peters and Robert H. Waterman, *In Search of Excellence: Lessons from America's Best-Run Companies* (New York: Harper & Row, 1982).

71. In 1952, DuPont refused to abide by Section 7A of the 1933 National Industrial Recovery Act, under which companies were required to negotiate with representative union locals.

72. On the 1960s, see Luc Boltanski and Eve Chiapello, *Le nouvel esprit du capitalisme* (Paris: Gallimard, 1999), 101.

73. Charles M. A. Stine, "The Engineer in a Free Economy" (HOBSO lecture, 1938), *Chemical Engineering Progress* 43, 9 (September 1947), HML.

74. Ibid.

75. Greenewalt, interview 1, 18.

76. Jasper Crane Papers, HML, ser. 2, box 1030.

77. Ibid.

78. F. C. Evans, letter to Jasper Crane, HML, ser. 2, box 1030.

79. HOBSO, lecture, 1938, HML.

80. Greenewalt, interview 1, 23.

81. Boltanski and Chiapello, *Nouvel esprit du capitalisme,* 95.

82. Ibid., 45.

83. See, e.g., John Woodhouse, interview 1, 1.

84. Cited by MacMullin in *Odyssey of a Chemical Engineer,* 75.

85. "Recruiting Personnel," HML, Acc. 1610.

86. These figures are based on employee rosters for the year 1940 and on "Women Workers of Du Pont," *Du Pont Magazine,* December 1940.

87. J. O. Maloney, "Doctoral Work in Chemical Engineering in the United States from the Beginning to 1960," in Furter, ed., *Century of Chemical Engineering,* 217.

88. ROTC was well entrenched in departments of chemical engineering at colleges and universities, but there are no statistics on how many DuPont engineers participated in it.

89. Greenewalt, interview 1, 1. On SAC and ROTC, see Noble, *America by Design*, 217–26.

90. Hounshell and Smith, *Science and Corporate Strategy*, 296. In *The Physicists: the History of a Scientific Community in Modern America* (New York: Knopf, 1977), 211, Daniel Kevles indicates that the industrial laboratories of large companies were widely tainted with antisemitism at the time.

91. Zunz, *Amérique en col blanc*, 119.

92. Chandler, *Strategy and Structure*, 136–37.

93. Frederic A. Wardenburg, letter to the "A"-bonus committee, HML, Acc. 1231, box 1028.

94. *Du Pont Magazine*, anniversary issue, 1927, 47.

95. Roger Williams interview, 4.

96. Cheape, *Strictly Business*, 112–13.

97. Arthur E. Morgan, cited by Roy Talbert Jr., *FDR's Utopian: Arthur Morgan of the TVA* (Jackson: University Press of Mississippi, 1987).

98. Daniel W. Mead, "The Engineer and His Education," in Jackson and Jones, eds., *Profession of Engineering*, 28.

99. C. Wright Mills, *White Collar: The American Middle Classes* (New York: Oxford University Press, 1951). On this subject, see also Zunz, *Making America Corporate*, introduction.

100. Crawford Greenewalt, "The Culture of the Businessman," *Du Pont Magazine*, April–May 1957, 13. Greenewalt is, of course, alluding to the stereotype in Sloan Wilson's famous book *The Man in the Gray Flannel Suit* (New York: Simon & Schuster, 1955).

101. Greenewalt, interview 2, 13.

FOUR: The Forgotten Engineers of the Bomb

1. The most detailed accounts are found in the two official histories of the Manhattan Project and the AEC: Vincent C. Jones, *Manhattan, the Army and the Atomic Bomb* (Washington, D.C.: Center of Military History, U.S. Army, 1985), which was commissioned by the Army and for the most part concentrates on the military; and Richard G. Hewlett and Oscar E. Anderson Jr., *The New World, 1939/1946*, vol. 1 of *A History of the United States Atomic Energy Commission* (University Park: Pennsylvania State University Press, 1962). See also the popular history by Richard Rhodes, *The Making of the Atomic Bomb* (New York: Simon & Schuster, 1986), even though this book essentially concentrates on the building of the bomb at Los Alamos rather than on the production of fissile materials. The memoirs I am using here are Arthur H. Compton: *Atomic Quest: A Personal Narrative* (New York: Oxford University Press, 1956), and Groves, *Now It Can Be Told*. To these should be added the biography of Leslie Groves by Stanley Goldberg, *Fighting to Build the Bomb: The Private Wars of Leslie Groves* (New York: Steerforth, 1995).

2. Hounshell and Smith, *Science and Corporate Strategy*, 338–46.

3. Hughes, *American Genesis*, 353–442, "Manhattan District."

4. The DuPont archives are in HML, Acc. 1957, Atomic Energy Division Records,

ser. 1, subser. A, B, C, D, and E. The archives of the U.S. War Department have been analyzed by V. C. Jones, *Manhattan,* and Hewlett and Anderson, *New World.*

5. Einstein's letter is reproduced in the annex to V. C. Jones, *Manhattan,* 609–10.

6. In 1939, Sachs had taken Niels Bohr's seminar at the Institute for Advanced Studies at Princeton, in which Bohr had informed the American physicists about the latest European advances in this field, in particular, the uranium fission achieved in Germany a few weeks earlier.

7. On the Third Reich's nuclear program, see Thomas Powers, *Heisenberg's War: The Secret History of the German Bomb* (New York: Knopf, 1993). Compton talks about the American physicists' worry about this matter in *Atomic Quest,* 23–24.

8. V. C. Jones, *Manhattan,* 12, and Hewlett and Anderson, *New World,* 14ff. Uranium was hard to come by; prior to the 1940s, there had not been not much demand for it. The United States had access to low-grade uranium ore from mines in Colorado and Utah, as well as to better-quality ore from a mine at Great Bear Lake in Canada, and the Union Minière du Haut Katanga supplied 1,200 tons of pitchblende from Shinkolobwe in the Belgian Congo. After they had landed in Europe, the Americans hunted down uranium in Belgium, Germany, and France. In October 1944, they confiscated 31 tons warehoused in Toulouse.

9. These advances were reported in detail in the Smyth Report, issued by the U.S. government in 1945 and subsequently published as Henry DeWolf Smyth, *Atomic Energy for Military Purposes: The Official Report on the Development of the Atomic Bomb Under the Auspices of the United States Government, 1940–1945* (Princeton, N.J.: Princeton University Press, 1945, 1947).

10. See Ruth Levine Sime, *Lise Meitner: A Life in Physics* (Berkeley: University of California Press, 1996).

11. V. C. Jones, *Manhattan,* 40–46.

12. Stone & Webster and M. W. Kellogg's role was what military jargon terms AME, for architect-engineer-manager.

13. "Initial Approach to Du Pont," HML, History TNX, Acc. 1957, ser. 1, A, box 1.

14. Groves, *Now It Can Be Told,* 95–96.

15. Compton, *Atomic Quest,* 109.

16. Greenewalt, interview 3, 23.

17. "Initial Approach," cited n. 13 above.

18. "Du Pont Assistance in Design of Semi-Works Facilities," HML, History TNX, Acc. 1957, ser. 1, A, box 1, p. 2.

19. Letter contract W–7412, HML, Acc. 1957, ser. 1, subser. B, "Contracts and Related Correspondence."

20. "Request for Evaluation of Feasibility of Large-Scale Production," in "Initial Approach," cited n. 13 above, 5.

21. Ibid.

22. "Preliminary Report on Feasibility," in "Initial Approach," cited n. 13 above, 10.

23. Greenewalt, interview 3, 7, indicates that this committee was his idea.

24. "Appointment of OSDR Committee to Review Alternative Processes," in "Initial Approach," cited n. 13 above, 12.

25. Hughes, *American Genesis,* 226.

26. Fermi's first chain reaction is recounted in a great many narratives, among them those of the participants, such as Compton, *Atomic Quest,* 141–45. In particular, this

experimentation made it possible to determine the factor k, which equals the number of neutrons at the moment t divided by the number neutrons at the moment $t-1$. The higher the k, the more exponential the chain reaction becomes (it is then called "supercritical"). If $k < 1$, more neutrons are absorbed than given off, and the reaction is smothered (in this case, it is called "subcritical").

27. Lewis Committee, "Conclusions of Reviewing Committee," 4 December 1942, National Archives, Administrative Files, General Correspondence, 334 (see V. C. Jones, *Manhattan*, 104–5). The report was written on the train as the members of the committee traveled from San Francisco to Chicago in late November or early December. At that point, they had already chosen the plutonium project, and the successful chain reaction was bound to confirm this decision. The committee's visit at Berkeley was "fascinating," according to Greenewalt, but its members were somewhat alarmed by Ernest Lawrence's overly optimistic statements, which were systematically toned down by J. Robert Oppenheimer.

28. Contract W–7412 eng–1, retroactively dated 1 December 1942, HML, Acc. 1957, ser. 1, subser. B, "Contracts and Related Correspondence."

29. Lewis Committee, "Conclusions of the Reviewing Committee," cited n. 27 above.

30. Groves, *Now It Can Be Told*, 46–59, and "Considerations of Government Request to Undertake Large-Scale Production," in "Initial Approaches," cited n. 13 above. The decision to forgo profits was stressed in a letter to stockholders of 13 August 1945, HML, Acc. 1957, *Stockholder Bulletin*.

31. S. L. Abrams, head of the legal department, letter to Charles Stine, presented to the finance board (which met just before the Executive Committee) on 21 December 1942, HML, Acc. 1957, ser. 1, subser. B, "Contracts and Related Correspondence."

32. "Letter dated December 16, 1942, from Dr. Vannevar Bush to President Roosevelt." Cited in V. C. Jones, *Manhattan*, 106.

33. The secret letter was marked: "In Order to Understand Thoroughly the Reasons and the Circumstances Lying Behind this Contract, It is Necessary to Consider with this Contract the Following Documents." These were the contents of the letter, a detailed account of the negotiations—both preliminary and legal—and the various technical memoranda. "Negotiations and Execution of Definitive Contracts," HML, Acc. 1957, ser. 1, subser. B, "Contracts and Related Correspondence."

34. *The Du Pont Company's Part in the National Security Program* (brochure published by DuPont in 1945), HML, Acc. 1410, box 2.

35. Lammot du Pont, Resolutions Committee, National Association of Manufacturers, New York, September 1942, HML, Lammot du Pont, personal files.

36. This information was subsequently invalidated when Compton and Seaborg confirmed the fissile character of plutonium; but what is important here is that in late 1942, the manufacturing of this new element was still hypothetical.

37. *Du Pont Magazine*, November–December 1962, HML.

38. Greenewalt, interview 3, 18.

39. Ibid., 28.

40. Hounshell and Smith, *Science and Corporate Strategy*, 79.

41. "Security, Disclosure of Information," in "Initial Approaches," cited n. 13 above, 56.

42. "Preliminary Organization and Work," in "Initial Approaches," cited n. 13 above, 16.

43. Greenewalt, interview 3.

44. Dale F. Babcock, "Du Pont and Nuclear Energy," 22 June 1982, HML, Acc. 1957, ser. 2, box 7.

45. This percentage is taken from V. C. Jones, *Manhattan*, 199, but I have been unable to verify it.

46. The number of DuPont employees on loan to the University of Chicago gradually increased until March 1944, when there were 392 of them. These employees were paid by the university, which was reimbursed by DuPont, which in turn was reimbursed by the government. This complicated procedure was used in order to make sure that the employees did not lose their benefits (e.g., salaries, bonuses, and pensions). "Loan of Personnel to University of Chicago," in "Initial Approaches," cited n. 13 above, 23.

47. Greenewalt, interview 3, 5.

48. Ibid., 21.

49. Ibid., 8.

50. This typology is taken from the first textbooks of nuclear engineering published after the war: S. Glasstone and M. C. Edlund, *The Elements of Nuclear Reactor Theory* (Princeton, N.J.: Van Nostrand, 1952); R. Stephenson, *Introduction to Nuclear Engineering* (New York: McGraw-Hill, 1954); and Arthur R. Foster and Robert L. Wright Jr., *Basic Nuclear Engineering* (Boston: Allyn & Bacon, 1968).

51. Greenewalt's Manhattan Project Diary begins on 16 December 1942 with the signing of the contract between DuPont and the Army. This is a technical and scientific notebook. "Unclassified Portions of C. H. Greenewalt's Notes," HML, Acc. 1889.

52. *Better Living*, November–December 1962, 18. HML, Acc. 1463, box 1.

53. "Supplement Number One to Letter Contract W–7412 eng 1," 4 January 1943, HML, Acc. 1957, ser. 1, subser. B, "Contracts and Related Correspondence."

54. Resolution of the Executive Committee, 15 January 1943, HML, Acc. 1957, ser. 1, subser. B, "Contracts and Related Correspondence."

55. Compton, *Atomic Quest*, 150–52ff.; Groves, *Now It Can Be Told*, 68–69.

56. In May 1945, the University of Chicago, no longer wishing to manage the pilot plant, transferred this responsibility to the Montsanto Chemical Company, which created a special department to do so. See Hewlett and Anderson, *New World*, 627.

57. Greenewalt, Manhattan Project Diary, 8, 9, 12, 15, 16, and 20 January 1943; Hewlett and Anderson, *New World*, 190–93.

58. The plant was officially called the Clinton Engineering Works (CEW) until the creation of the Atomic Energy Commission in 1947, when the place-name Oak Ridge came to be used. Groves, *Now It Can Be Told*, 25–26.

59. "Clinton Field Construction," HML, History TNX, 33.

60. Wheeler had been Niels Bohr's assistant when the latter demonstrated that only ^{235}U was fissile.

61. Hewlett and Anderson, *New World*, 168–69.

62. Greenewalt, interview 3, 13 and 27.

63. Hewlett and Anderson, *New World*, 203.

64. Reactors for military use do not in fact need the heat generated by the reaction, unlike reactors for civilian use: the latter use that heat to produce electricity; but on the other hand civilian engineers have no use for the irradiated uranium produced by the reaction.

65. "Selection of Water Cooled Pile," HML, History TNX, 26–27. Water is an excellent coolant, but it absorbs many neutrons and it is corrosive. Helium has the disadvan-

tage of having to be used under pressure, and this made for extremely tricky problems of construction and maintenance.

66. HML, History TNX, 22.

67. Alain Dewerpe, "L'exercice des conventions," in Bernard Lepetit, ed., *Les formes de l'expérience* (Paris: Albin Michel, 1995), 114.

68. Greenewalt, interview 3, 25.

69. Richard Feynman, *"Surely you're joking, Mr. Feynman!" Adventures of a Curious Character* (1985; New York: Bantam Books, 1989), 106.

70. Greenewalt, interview 3, 1.

71. Meeting of the metallurgical laboratory of 28 December 1942: Greenewalt, Manhattan Project Diary, 28 December 1942.

72. Fermi quoted by Greenewalt, interview 3, 24.

73. Greenewalt, interview 3, 2.

74. Letter from Sam Allison to Williams, photocopied and inserted in Greenewalt's Manhattan Project Diary, 13 March 1944.

75. Jean-François Lyotard, *Le différend* (Paris: Minuit, 1983), 9.

76. "Site Considerations—Selection of Hanford Site," HML, History TNX, 17, and V. C. Jones, *Manhattan*, 109.

77. "General Site Requirements," HML, History TNX.

78. "Hanford Field Construction," HML, History TNX, 34–35.

79. "Design of Richland Housing Facilities," HML, History TNX, 38–39.

80. Ibid., 36.

81. Greenewalt, interview 3, 29–30.

82. "Preliminary Design, Water Cooled Pile—Corrosion," HML, History TNX, 28–29.

83. This episode is told in V. C. Jones, *Manhattan*, 221.

84. See, e.g., Foster and Wright, *Basic Nuclear Engineering*, "Rates of Change of Poison Nuclei in a Reactor," 243.

85. Greenewalt, interview 3, 26.

86. Hewlett and Anderson, *New World*, 308.

87. Hounshell, "Du Pont and the Management of Large-Scale Research and Development," 25. Oddly, Hounshell does not make use of his interview with Greenewalt, which puts this anecdote into context (26).

88. "Old Marse George," reproduced in Compton, *Atomic Quest*, 192–93.

89. Greenewalt, interview 3, 35.

90. At the moment of release, a contraption propelled two blocks of ^{235}U placed at opposite ends of Little Boy's nose cone toward each other, whose combined energy instantly reached critical mass, triggering a chain reaction that liberated an amount of energy equivalent to 12,000 tons of TNT. Neutron reflectors further increased the amount of fission obtained, so that only 15 kg of ^{235}U was needed.

91. On these questions of design, see Hewlett and Anderson, *New World*, and Rhodes, *Making of the Atomic Bomb*, 702–3.

92. On the Soviet nuclear program and related espionage activities, see Richard Rhodes, *Dark Sun: The Making of the Hydrogen Bomb* (New York: Simon & Schuster, 1995), 267–68. The first Soviet reactor ("F1") to start a chain reaction, on 25 December 1946, was an exact copy of one of the Hanford reactors (except for the vertical loading of the uranium rods). It is not known who transmitted this information to Beria and

Kurchatov, but it is likely that Hanford had another "mole," in addition to Klaus Fuchs, who worked mostly at Los Alamos. See also David Holloway, *Stalin and the Bomb* (New Haven, Conn.: Yale University Press, 1994).

93. On the atomic explosion at Nagasaki and its consequences, see Gar Alperovitz, *The Decision to Use the Atomic Bomb and the Architecture of an American Myth* (New York: Knopf, 1995), which is filled out by Barton Bernstein: "A Postwar Myth: 500,000 American Lives Spared," *Bulletin of Atomic Scientists* 42, 6 (1986), and "Seizing the Contested Terrain of Early Nuclear History: Stimson, Conant and their Allies Explain the Decision to Use the Atomic Bomb," *Diplomatic History* 17, 1 (1993), and by the articles published in *Diplomatic History* 19, 2 (Spring 1995): 197–365. For testimonies of survivors, see, e.g., Masji Ibuse, *Black Rain* (Tokyo: Kodansha International, 1969); Makoto Oda, *The Bomb* (New York: Kodansha International, 1990); and Kyoko Selden and Mark Selden, eds., *The Atomic Bomb: Voices from Hiroshima and Nagasaki* (Armonk, N.Y.: M. E. Sharpe, 1989). In a third category of books are those concerning the impact of the bomb on our contemporary mental images: Paul Boyer, *By the Bomb's Early Light: American Thought and Culture at the Dawn of the Atomic Age* (1985; repr. with new preface, Chapel Hill: University of North Carolina Press, 1992); H. Bruce Franklin, *War Stars: The Superweapon and the American Imagination* (New York: Oxford University Press, 1988); and Spencer R. Weart, *Nuclear Fear: A History of Images* (Cambridge, Mass.: Harvard University Press, 1988).

94. The power of "Little Boy" and "Fat Man" is expressed in kilotons, that is to say, in the equivalent of a thousand tons of TNT. This notion measures the force of the explosion and not its destructive power by comparison with regular bombs. Destroying Hiroshima and Nagasaki by conventional bombing would have required 2,100 and 1,200 tons of bombs, or two raids by 210 and 120 B-29 bombers, respectively.

95. Greenewalt, interview 3, 25. The physicists, notably Lawrence at Berkeley, now began to build cyclotrons (the first particle accelerators) with the help of subcontractors, for whose work the scientists were responsible. On the cyclotrons, see Robert Seidel, "The Origin of the Lawrence Berkeley Lab," in Galison and Hevly, eds., *Big Science,* 21–45.

96. See Holloway, *Stalin and the Bomb,* and Nikolaus Riehl and Frederick Setz, *Stalin's Captive: Nikolaus Riehl and the Soviet Race for the Bomb* (Philadelphia: American Chemical Society and Chemical Heritage Foundation, 1996). This latter work transcribes the testimony of a German atomic physicist who, transferred to the Soviet Union in 1945, was made to work for the Soviet nuclear program for ten years.

97. Rhodes, *Dark Sun,* 266–67, cites a Russian physicist, Sergei Frish.

98. Greenewalt, interview 3, 44.

99. Greenewalt, interview 1, 45.

100. James O. Maloney, "Chemical Engineering Education at the University of Kansas, 1895–1988," in John Parascandola and James Whorton, eds., *Chemistry in Modern Society* (Washington, D.C.: American Chemical Society, 1983), 332.

101. "Guests for Premiere of 'The Beginning or the End,' 5 March 1947, Washington, Loew's Aldine Theater," HML, Acc. 1410, box 34.

102. The reviews of the film were uniformly negative, essentially because of the obvious distortions in the screenplay—such as the claim that the inhabitants of Hiroshima were warned by leaflets ten days before the bombardment.

103. James G. Vail, in *Chemical Engineering Progress,* February 1947, 5.

104. *Chemical Engineering Progress,* March 1947, 24.

105. Albert B. Newman (president of the AIChE for the year 1948), "Your Institute in 1948," *Chemical Engineering Progress* 44 (January 1948): 1.

106. Engineers' Joint Council and Bureau of Labor Statistics, "On the Economic Status of Engineers," 1947. The survey examined 46,110 engineers in all specialties.

107. Cheape, *Strictly Business,* 226.

108. See Selznick, *TVA and the Grass Roots,* 1984.

109. David Lilienthal, *Big Business: A New Era* (New York: Harper, 1953); Potter, *People of Plenty.*

FIVE: The Heyday and Decline of Chemical Engineering

1. These figures are taken from Galambos and Pratt, *Rise of the Corporate Common-wealth,* 101 and 134.

2. James Forrestal, 10 May 1943, cited by Michael Sherry, *Preparing for the Next War: American Plans for Post-War Defense, 1941–45* (New Haven, Conn.: Yale University Press, 1977), 33.

3. Ibid., ix.

4. Boyer, *By the Bomb's Early Light,* and Weart, *Nuclear Fear.*

5. Eric Johnston, *America Unlimited* (Garden City, N.J.: Doubleday, Doran, 1944); Galambos and Pratt, *Rise of the Corporate Commonwealth,* 128.

6. Johnston, *America Unlimited,* 152.

7. See esp. John Kenneth Galbraith, *American Capitalism: The Concept of Countervailing Power* (Boston: Houghton Mifflin, 1952) and *The Affluent Society* (Boston: Houghton Mifflin, 1958). Galbraith continued in the tradition inaugurated by the Progressive historians a half century earlier when he pointed out that, contrary to the assertion of the chairman of General Motors, what was good for the United States was not necessarily good for General Motors—and vice versa.

8. If from now on this study focuses on the managerial and bureaucratic functions of the chemical engineers, it is because the engineers whose careers we have followed had become managers and left the actual engineering work to a younger generation. Whereas technological and scientific aspects had occupied a central position in their work during the 1930s and the early 1940s, they now spent most of their time on supervisory and administrative functions. If chemical engineering as such seems to fade away in the following pages, it is precisely because I do not wish to adopt a "substantialist" concept of this professional group, for such a concept would once and for all define the delimiting criteria of the group (on the basis of a technical division of labor), thereby reifying its representatives without considering the evolution some of them may have undergone.

9. See Paul Edwards, *The Closed World: Computers and the Politics of Discourse in Cold War America* (Cambridge, Mass.: MIT Press, 1996).

10. *Grand Rapids Herald,* 3 October 1945. A detailed review of the press is found in HML 1410, box 50.

11. "The Lighter Side of Nylon," HML 1410, box 50.

12. Los Angeles man, letter to DuPont, HML 1410, box 50.

13. Henry B. du Pont, "Du Pont in the Postwar Period" (address to the annual meeting of the Rayon Department, 14 April 1944), HML 1410, box 61-A.

14. Greenewalt, Manhattan Project Diary, 27 July 1944. The page includes a memorandum to Roger Williams about this question.

15. Greenewalt, interview 3.

16. Greenewalt, Manhattan Project Diary, 2 December 1943.

17. See Boyer, *By the Bomb's Early Light.*

18. Stine to the Executive Committee, 12 February 1945, ser. 2, Acc. 2, box 830.

19. Hewlett and Anderson, *New World,* 630.

20. Brian Balogh, *Chain Reaction: Expert Debate and Public Participation in American Commercial Nuclear Power, 1945–1975* (New York: Cambridge University Press, 1991), chap. 3.

21. Ibid., 48.

22. "Pension Plans in AEC Contracts," HML, Legal Department, 1957, ser. 2, box 5.

23. These debates can be followed in Richard Hewlett and Francis Duncan, *Atomic Shield, 1947–1952,* vol. 2 of *A History of the United States Atomic Energy Commission* (University Park: Pennsylvania State University Press, 1969).

24. Margaret Smith Stahl, "Split and Schism, Nuclear and Social" (Ph.D. diss., University of Wisconsin, 1946), 257ff.

25. Rhodes, *Dark Sun,* 363.

26. Ibid.; Balogh, *Chain Reaction;* Hewlett and Duncan, *Atomic Shield;* Herbert York, *The Advisors: Oppenheimer, Teller, and the Superbomb* (San Francisco: W. H. Freeman, 1975).

27. Rhodes, *Dark Sun,* 420.

28. See Marcelle Knaak, "Post World War II Bombers, 1945–1973," *Encyclopedia of US Air Force Aircraft and Missile Systems,* vol. 2 (Washington, D.C.: Office of Air Force History, 1988). On the Strategic Air Command, see Eric Markusen and David Kopf, *The Holocaust and Strategic Bombing: Genocide and Total War in the Twentieth Century* (Boulder, Colo.: Westview Press, 1995). And see also David Alan Rosenberg, "American Atomic Strategy and the Hydrogen Bomb Decision," *Journal of American History* 66, 1 (June 1979): 62–87.

29. "AEC Construction Programs, 1947–1950," National Archives, College Park, Md., RG 326.4.

30. The negotiations for the contract are detailed in "History of Negotiation of Contract AT (07-2)-1," HML, Acc. 1957, ser. 2, box 5.

31. Ibid.

32. "Atomic Energy Survey Committee and Visits to AEC Installations by the Military, Boards, Committees, 1947–1950," National Archives, College Park, Md., RG 326.1.

33. President Harry S Truman, letter of 25 July 1950, reproduced in William P. Babbington, *History of Du Pont at the Savannah River Plant* (Wilmington, Del.: DuPont, 1990).

34. Resolution of the Executive Committee, 26 July 1950.

35. U.S. Congress, Joint Committee on Atomic Energy, Hearings, 81st Cong., 2d sess., 1950.

36. "Salaries of Contractor Employees," National Archives, College Park, Md., RG 326.5.

37. These figures have been established on the basis of DuPont's annual reports to its shareholders.

38. Balogh, *Chain Reaction,* 68. On expertise, see Brian Balogh, "Reorganizing the

Organizational Synthesis: Federal-Professional Relations in Modern America," *Studies in American Political Development* 5 (Spring 1991): 119–72; Frank J. Newman, "The Era of Expertise: The Growth, the Spread and Ultimately the Decline of the National Commitment to the Concept of the Highly Trained Expert: 1945 to 1970" (Ph.D. diss., Stanford University, 1981); Thomas L. Haskell, ed., *The Authority of Experts: Studies in History and Theory* (Bloomington: Indiana University Press, 1984).

39. "General Advisory Committee Correspondence, 1947–1951," National Archives, College Park, Md., RG 326.1.

40. "Division of Production—Organization and Functions, 1947–1951," National Archives, College Park, Md., RG 326.2.

41. Hood Worthington, "Memorandum for Consideration by the General Advisory Committee," Atomic Energy Commission, GAC, meeting, 28 March 1947, in "General Advisory Committee Minutes of Meetings," vols. 1–4, National Archives, College Park, Md., RG 326.1.

42. "Inspection Tours and Visits by Members of the AEC, 1947–1951," National Archives, RG 326.1.

43. Roger Williams, "Report on Hanford Inspection," 14 July 1947, National Archives, College Park, Md., RG 326.1.

44. Atomic Energy Commission, "General Advisory Committee Minutes of Meetings," vols. 1–4, National Archives, College Park, Md., RG 326.1.

45. On the JRDB, see Allen A. Needell, "From Military Research to Big Science: Lloyd Berkner and Science-Statesmanship in the Postwar Era," in Galison and Hevly, eds., *Big Science*, 290–311.

46. "Nuclear Propulsion of Aircraft Program (NEPA)," National Archives, College Park, Md., RG 326.5.

47. "Air Force Operational War Plans," 1948–1949, National Archives, College Park, Md., RG 326.5.

48. A multitude of more or less serious plans had been hatched in the general staffs. Only Hyman G. Rickover, whose perseverance was legendary, succeeded, building the first nuclear-powered submarine, the USS *Nautilus*, launched in 1954. See Richard Hewlett and Francis Duncan, *Nuclear Navy, 1946–1962* (Chicago: University of Chicago Press, 1974).

49. Hewlett and Duncan, *Atomic Shield*, 72–73.

50. "Committee on Atomic Energy, Joint Research and Development Board, 30 July 1947," National Archives, RG 326.1.

51. Haynes, *Chemical Who's Who.*

52. Hewlett and Duncan, *Atomic Shield*, 157.

53. "Security, Coordination Between AEC and MLC, 1947–1949," National Archives, College Park, Md., RG 326.2.

54. "Request for Information by MLC," National Archives, College Park, Md., RG 326.1.

55. "Report by Carpenter to Lilienthal, 23 April 1948," National Archives, College Park, Md., Correspondence, Persons, RG 326.2.

56. Donald Carpenter, "Preliminary Draft of Report of the United States Atomic Energy Commission by the Industrial Advisory Group," 29 May 1948, National Archives, College Park, Md., RG 326.2. Carpenter wrote this report as a member of the Industrial Advisory Group of the AEC.

57. "AEC and MLC," National Archives, RG 326.1.

58. Hewlett and Duncan, *Atomic Shield,* 340.

59. Ibid., 160, cites Carpenter's "Report of Conversation with Lilienthal and Wilson" of 12 April 1948.

60. See McNamara, *In Retrospect,* chap. 1, "My Journey to Washington." John K. Galbraith evidently suggested to Kennedy that he appoint McNamara. The explicit aim was to reorganize the Department of Defense based on the systems analysis developed during World War II to plan military operations and favored by the RAND Corporation, which McNamara and his assistant Charles Hitch subsequently did. Donald Carpenter's reports were grounded in the prewar management tradition, however, rather than in systems analysis, which makes much use of mathematics.

61. See Thomas P. Hughes, *Rescuing Prometheus* (New York: Pantheon Books, 1998), esp. chap. 3.

62. HML, Acc. 1957, ser. 1, subser. B, box 1. This last survey, undertaken at the request of the AEC in December 1948, was brought to a close in March 1949 with a detailed report on the American nuclear capabilities on the eve of the explosion of the first Soviet atomic bomb.

63. HML, AED Division, ser. 1, subser. B, box 1, and National Archives, Correspondence, Persons, RG 326.2.

64. See Seymour Melman, *The Permanent War Economy: American Capitalism in Decline* (New York: Simon & Schuster, 1974).

65. Robert Griffith, "Dwight D. Eisenhower and the Corporate Commonwealth," *American Historical Review* 87, 1 (1982): 87–122, and "President Eisenhower Warns of the MIC, 1961," in Robert Griffith, ed., *Major Problems in American History Since 1945* (Lexington, Mass.: D. C. Heath, 1992), 166–67.

66. This anecdote is told by Hewlett and Duncan, *Atomic Shield.*

67. "History of Negotiation of Contract AT (07-2)-1," HML, Acc. 1957, ser. 2, box 5.

68. Greenewalt, letter to Truman, 17 October 1950, HML, Acc. 1957, ser. 2, box 5.

69. On nylon in fashion, see Susannah Handley, *Nylon: The Story of a Fashion Revolution* (Baltimore: Johns Hopkins University Press, 1999).

70. Rhodiaceta was a branch of Rhône-Poulenc, today the world's second-largest producer of nylon.

71. "Agreement Between Du Pont de Nemours and the Chemstrand Corporation," 2 June 1951, HML, Acc. 1850.

72. Hounshell and Smith, *Science and Corporate Strategy,* 440–41. ETF stood for Engineering and Textile Fibers Departments. This process was called "Blue Sky Nylon Technology."

73. H. F. Brown, letter to Carleton A. Shugg, acting general manager of the AEC. On the general organization of the AED, see "Savannah River Organization, 1950," National Archives, RG 326.1, and HML, ser. 2.

74. The Department of Explosives was abolished in 1972 and its activities were taken over by the Polymer Intermediate Department. Thus the AED was administratively connected with the Department of Petrochemicals.

75. "Expenditures Fiscal Year Ending June 30, 1951, 1952, 1953, 1954," HML, AED, Acc. 1957, ser. 2, box 13.

76. "AEC Relationship with Its Contractors," vols. 1–3, National Archives, RG 326.5.

77. HML, ser. 2, subser. A.

78. "Recruitment of Technical Personnel for Operations," HML, Acc. 1957, ser. 2, box 5.

79. Ibid., and ser. 2, subser. B.

80. "Recruitment of Technical Personnel," cited n. 78 above.

81. "Executive Committee Request to Heads of Departments to Make Qualified Men Available," 2 September 1950, HML, AED, Acc. 1957, ser. 2, box 5.

82. Ibid.

83. "The general policy is that employees are not to be drafted or assigned by the management, but that, with due regard to the importance of other work, qualified employees are to be offered the opportunity for voluntary transfer to the Atomic Energy Division."

84. Babcock entry in William Haynes, ed., *Chemical Who's Who: Biography in Dictionary Form of the Leaders in Chemical Industry, Research, and Education*, 4th ed. (New York: Lewis Historical Publishing Co., 1956).

85. "Letter of Offer, Standard Model; Salaries of Contract Employees," National Archives, RG 326.5.

86. "Security Clearance of Personnel—Policy and Procedures," National Archives, RG 326.5.

87. "Scientific Personnel Policy of Contractors," National Archives, RG 326.5.

88. Babbington, *History of Du Pont at the Savannah River Plant*, 13–14.

89. See Melman, *Permanent War Economy.*

90. See Ann Markusen, Peter Hall, Scott Campbell, and Sabina Deitrick, *The Rise of the Gunbelt: The Military Mapping of Industrial America* (New York: Oxford University Press, 1991).

91. See Edward Teller, ed., *Fusion* (New York: Academic Press, 1981), and Friedwardt Winterberg, *The Physical Principles of Thermonuclear Explosive Devices* (New York: Fusion Energy Foundation, 1981).

92. Ferdinand G. Brickwedde, "Harold Urey and the Discovery of Deuterium," *Physics Today*, 1982, 9.

93. On Ulam, see his memoirs, Stanislaw Ulam, *Adventures of a Mathematician* (New York: Scribner, 1976).

94. This Ulam-Teller scheme is well known to nuclear physicists today, but for a long time, the Americans kept it secret, so that French and British scientists had to discover it for themselves. See Rhodes, *Dark Sun*, chap. 23.

95. That is why it is rather easy to distinguish civilian from military facilities. If the uranium rods are within easy reach and if the permanent loading and unloading devices are visible, the production is for military ends. There are also mixed facilities.

96. Babbington, *History of Du Pont at Savannah River*, 41–43.

97. Ibid.

98. On the manufacture of increasingly powerful fissile materials, see Richard Hewlett and Jack M. Holl, *Atoms for Peace and War* (Berkeley: University of California Press, 1989), chap. 14.

99. "Nuclear Notebook," *Bulletin of Atomic Scientists*, December 1993, 57.

100. Today the Nuclear Regulatory Agency and the Energy Research and Development Administration, two branches of the Department of Energy, have replaced the AEC, which was dissolved in 1974.

101. Babbington, *History of Du Pont at the Savannah River Plant*, 217–18.

102. HML, Acc. 1957, ser. 2, box 5.

103. Over the next few years, this was done only in a marginal manner, essentially because only a few countries adopted the technology developed by DuPont (i.e., reactors moderated by heavy water), among them Canada. DuPont therefore worked with the Canadian Nuclear Safety Commission, but was never able to attain the levels reached by General Electric and Westinghouse.

104. R. C. Gunness, *Chemical Engineering Progress* 47, 7 (June 1951): 331.

105. W. Laid Stabler, "Tomorrow's Manager," HML.

106. This argument is also advanced in an article in *Du Pont Magazine,* November–December 1960, 6–7.

107. Laure M. Sharp, *Manpower Resources in Chemistry and Chemical Engineering,* U.S. Department of Labor, Bureau of Labor Statistics Bulletin 1132 (Washington, D.C.: Government Printing Office,1953). This study is divided into two parts, with the first devoted to the chemists and the second to the chemical engineers.

108. United States Bureau of the Census, 1950 Census.

109. Sharp, *Manpower Resources,* table A–31.

110. Ibid., table A-34.

111. Ibid., table A-37.

112. Ibid., table A-39.

113. Ibid., table 40.

114. Ibid., table A-44.

115. Ibid., table 7.

116. *Chemical Engineering Progress* 48, 8 (July 1951): 420–21.

117. J. O. Maloney, "Doctoral Thesis Work in the US," in Furter, ed., *Century of Chemical Engineering,* 213.

118. Ibid.

119. Lawrence Grayson, "A Brief History of Engineering Education in the United States," *Engineering Education* 67 (December 1977): 260.

120. Interview of Stuart Churchill by the author. I thank him for speaking to me.

121. Cole Coolidge, "Organization of Du Pont Research," 23 June 1952, HML, Acc. 1850.

122. Greenewalt, interview 1.

123. "The Profs Come to Watch," *Du Pont Magazine,* August–September 1958, 16–17.

124. Ibid.

125. H. C. Weber, "The Improbable Achievement: Chemical Engineering at MIT," in Furter, ed., *History of Chemical Engineering,* 89.

126. See also Zunz, *Amérique en col blanc,* for the situation at other companies.

127. *Time,* 16 April 1951, 94ff.

128. Frederick W. Taylor, *Scientific Management* (New York: Harper, 1911), 7.

129. See Thorstein Veblen, *The Engineers and the Price System* (New York: B. W. Huebsch, 1921).

130. On the Technology Movement, see Thomas P. Hughes, ed., *Changing Attitudes Toward American Technology* (New York: Harper, 1975), 297ff.

131. In this connection, see Selznick, *TVA and the Grass Roots,* and Talbert, *FDR's Utopian.*

132. See Balogh, "Reorganizing the Organizational Synthesis."

133. John F. Kennedy, address at Yale University, 11 June 1962, www.jfklibrary.org/j061162.htm (accessed 10 February 2006).

134. C. H. Greenewalt, "The Responsibilities of Business Leadership," read by Lammot du Pont Copeland (DuPont's eleventh president, great-great-grandson of Eleuthère Irénée du Pont), 13 September 1961, HML, Acc. 1410, box 39.

135. Richard Hofstadter, *Social Darwinism in American Thought* (Philadelphia: University Press of Pennsylvania, 1944). In this connection, see Susan Stout Baker, *Radical Beginnings: Richard Hofstadter and the 1930s* (Westport, Conn.: Greenwood Press, 1985), and Daniel Joseph Singal, "Beyond Consensus: Richard Hofstadter and American Historiography," *American Historical Review* 89 (1984): 976–1004.

136. Richard Hofstadter, *The Age of Reform: From Bryan to FDR* (New York: Knopf, 1955), 223. A good analysis is found in David Noble's *The End of American History: Democracy, Capitalism and the Metaphor of Two Worlds in Anglo-American Historical Writing* (Minneapolis: University of Minnesota Press, 1985).

137. On the labor movement during World War II, see Nelson Lichtenstein, *Labor's War at Home: The CIO in World War II* (New York: Cambridge University Press, 1982).

138. David Noble, *Forces of Production: A Social History of Industrial Automation* (New York: Oxford University Press, 1984), 25.

139. *Du Pont Magazine*, April 1949.

140. See Dominique Janicaud, *La puissance du rationnel* (Paris: Gallimard, 1985).

141. Perrin Stryker, "Chemicals: The Ball Is Over," *Fortune*, October 1961.

142. Hounshell and Smith, *Science and Corporate Strategy*, 512.

143. "Discussions with General Managers on Company Performance and Organization," 16 July 1961, HML, Acc. 1814, box 3.

144. Michele Gerber, *On the Home Front: The Cold War Legacy of the Hanford Nuclear Site* (Lincoln: University of Nebraska Press, 1992), 5.

145. Robert Alvarez, "The Legacy of Hanford," *The Nation*, 18 August 2003, 1 (www.thenation.com/doc/20030818/alvarez [accessed 2 February 2006]). Alvarez was senior policy adviser to the U.S. secretary of energy from 1993 to 1999.

146. "Report of the Study on Du Pont's Research Reputation, by the Research Reputation Planning Committee, Public Relations Department," HML, Acc. 1410, box 18.

147. Ibid., 7.

148. Cited by Jacqueline Rémy, "1937–1978, les années nylon," *L'Express*, 20–26 November 1987, 121.

149. Ibid., citing a survey by Marc-Alain Descamps.

150. J. O. Maloney, "Doctoral Thesis Work in Chemical Engineering," in Furter, ed., *Century of Chemical Engineering*, 213.

151. National Science Foundation, *Science and Engineering Indicators—2000* (Arlington, Va.: NSF, 2000), www.nsf.gov/statistics/seind00 (accessed 2 February 2006).

Conclusion

1. Primo Levi, *The Periodic Table*, trans. Raymond Rosenthal (New York: Schocken Books, 1984).

2. Bernard Lepetit, "Histoire des pratiques, pratiques de l'histoire," in id., ed., *Les formes de l'expérience: Une autre histoire sociale* (Paris: Albin Michel, 1995), 12.

3. Olivier Zunz, "Recontrer l'histoire Américaine," in Jean Heffer and François Weil, eds., *Chantiers d'histoire américaine* (Paris: Belin, 1994), 436.

Essay on Sources and Historiography

The main archival sources used in this book are to be found in the National Archives in Washington, D.C., and at College Park, Maryland, and especially at the Hagley Museum and Library in Wilmington, Delaware (the birthplace of E. I. du Pont de Nemours and Company), which also holds the public archives of the Sun Oil Company, Remington Rand, Inc., and several other companies in addition to those of DuPont. I say public archives, for as historians of business know only too well, not all business archives are available to academic historians (e.g., the IBM archives).

Working with these sources presents its share of difficulties. DuPont did not systematically keep personnel files; lists of employees do not differentiate chemical engineers from other engineers until the end of the 1930s. The census also treats chemical engineers somewhat anonymously until World War II. Readers will thus find here neither flawless statistical series from the 1910s to the 1960s nor a representative sample of chemical engineers at DuPont, but only the collective biography of several dozen engineers whose careers I have been able to retrace. There is no doubt that these men left a strong mark on their profession and their firm and had an impact on their country. Eschewing any preconceived definition, I follow their careers through the turns and twists of organizational charts and along the path of technical projects, piecing together a collective portrait of this uniquely influential socioprofessional group as best as the record allows.

Then, too, the stories of nylon and the Manhattan Project are rather daunting, for they each constitute a complex topic, already treated in numerous historical and journalistic accounts. They are rich in events and usually told as success stories. By the end of the 1940s, many books had already heralded nylon as one of the achievements of the "American century." And in the early 1960s, some of the principal actors of the Manhattan Project, General Leslie Groves and the physicist Arthur Compton among them, revealed closely held military secrets, while laying out their own exploits in epic breadth. Hence the importance of reassessing these stories from the vantage point of the professional lives of the heretofore neglected group of engineers I focus on.

This approach is also useful because it permits me to treat the story of DuPont from a historical perspective, rather than a managerial one, thus making a contribution to the "reorganization of the organizational synthesis" called for by the political historian Brian Balogh in "Reorganizing the Organizational Synthesis," *Studies in American Political Development* 5 (1991) 119–72. The history of the engineers presented here is an attempt to recreate not only the participants' agency but also the contexts in which they worked and refashioned their firms. I want to avoid the methodological presupposition of too explicit a strategy on the part of any single group of actors. Rather, it was a matter of "emerging strategies" made necessary by particular circumstances, as the management science expert Henry Mintzberg has shown in "Patterns in Strategy Formation," *Management Science,* May 1978, and *The Structuring of Organizations: A Synthesis of the Research* (Englewood Cliffs, N.J.: Prentice-Hall, 1979). See also Patrick Friedenson, "Les organisations: Un nouvel objet," *Annales. Histoire, Sciences Sociales* 44, 6 (November–December 1989): 1461–77.

To put it differently, companies adapted to the rising power of the federal government by accommodating its increasing interventionism. In this interventionist context, a new generation of engineers made its way into the higher echelons of companies and brought about a major transformation of America's political economy.

One might argue that a study focusing on a major company belongs to the great tradition of business history. This historiography is dominated by studies of great firms that have "kindled the historical imagination," as Nancy L. Green observes in *Ready to Wear and Ready to Work: A Century of Industry and Immigrants in Paris and New York* (Durham, N.C.: Duke University Press, 1997). Green and other authors, such as Louis Bergeron in France and Philip Scranton, Charles Sabel, Jonathan Zeitlin, and Michael Piore in the United States, have pleaded for increased attention to smaller industries and a less monolithic view of industrialization. See Charles F. Sabel and Jonathan Zeitlin, "Historical Alternatives to Mass Production: Politics, Markets and Technology in Nineteenth-Century Industrialization," *Past and Present* 108 (1985): 133–76; Sabel and Zeitlin, eds., *World of Possibilities: Flexibility and Mass Production in Western Industrialization* (Cambridge, Mass.: Harvard University Press, 1997); Philip Scranton, *Endless Novelty: Specialty Production and American Industrialization, 1865–1925* (Princeton, N.J.: Princeton University Press, 1997). However, although it is true that far from being an archaic mode of production doomed to disappear sooner or later, small business is a dynamic and flexible sector of modern capitalism, this does not mean that the history of large companies should be abandoned. A good example of such a study is provided by Olivier Zunz in *Making America Corporate, 1870–1920* (Chicago: University of Chicago Press, 1990). One needs perhaps to go beyond the functionalist and internalist model promoted by the business historian Alfred Dupont Chandler Jr., because functionalism tends to mask conflicts and disagreements by presenting overly linear and deterministic analyses in which every element is a predictable part of the whole. Moreover, the internalist perspective stresses the tripartite market-strategy-structure relationship without taking into consideration interactions with other organizations, society, and the body politic.

The Progressive Historiographical Tradition

Most of the giant companies that were to contribute so much to shaping the face of twentieth-century America arose and developed at the end of the nineteenth century, when firms like E. I. Du Pont de Nemours and Company, the United States Steel

Corporation, the General Electric Company, the American Telephone and Telegraph Corporation (AT&T), and soon thereafter Ford, General Motors, and many others, imposed their organizations, products, production methods, and logos on both the United States and the rest of the world. The spectacular rise of big business, often led by colorful captains of industry, has naturally captured the attention of researchers. In the first half of the century, the Progressive historians harshly criticized the dubious strategies of big business and sympathized with its opponents, whether labor activists, reform-minded intellectuals, or politicians. These historians (foremost among them Charles Beard, but also Matthew Josephson, who dubbed Jay Gould "Mephistopheles") stressed the opposition between the people, democracy, and the public interest, on the one hand, and special interests, on the other. In their eyes, the "robber barons" cared nothing for the common good, treated government as their private hunting preserve, and in general represented capitalism run amok. See Charles Beard, *The American Leviathan: The Republic in the Machine Age* (New York: Macmillan, 1930); Matthew Josephson, *The Robber Barons: The Great American Capitalists, 1861–1901* (1934; repr., San Diego: Harcourt Brace, 1995), and Marianne Debouzy, *Le capitalisme sauvage, 1860–1900* (Paris: Seuil, 1972). At the beginning of the twentieth century, it seemed as though Congress and the White House had become mere branches of Standard Oil, and that the political will of the people and its representatives needed to be aroused against these unscrupulous businessmen.

Muckraking Progressive journalists exposed political corruption, fraud of every description, exploitation of children, and other dark and objectionable aspects of American business. In one of the first series of such articles, which appeared in *McClure's* magazine in 1902, Ida Tarbell focused on Standard Oil, revealing the methods John D. Rockefeller had employed in order to ensure his monopoly on petroleum and decrying the pollution caused by drilling and refining in the oil-producing areas of Pennsylvania and Ohio. See Ida M. Tarbell, *The History of the Standard Oil Company* (2 vols.; New York: McClure, Phillips, 1904).

The Progressives especially targeted the chemical industry, but from World War I to the early 1960s, critical studies of it were very rare. As the need for production was paramount, few Americans were willing to fight for the environment and against workplace accidents. The textbooks on industrial chemistry written in that period sang its praises and emphasized its great achievements, from aspirin to nylon and penicillin. Even though DuPont's management was vigorously attacked during the 1930s, the chemical industry as such was not targeted by this criticism.

Chemical companies commissioned books and radio broadcasts that depicted them in a highly favorable light, such as William S. Dutton's books *Du Pont: One Hundred and Forty Years* (New York: Scribner, 1942) and *Du Pont: Autobiography of a Scientific Enterprise* (New York: Scribner, 1952), and *Cavalcade of America*, a dramatized radio series sponsored by DuPont that was broadcast by CBS from 1935 to 1953. See also William Haynes, *American Chemical Industry: A History* (New York: Van Nostrand, 1954), a book commissioned by the American Chemical Society.

Progressivism, which primarily sprang from moral indignation, did not survive the

1940s and big business's concurrent transformation of itself. World War II and then the Cold War shifted the focus to the struggle against totalitarianism, whether Nazi or Soviet, and away from the big companies. Moreover, the latter seemed to have proven their worth when they transformed the United States into the "arsenal of democracy," while at the same time providing the American people with an inexhaustible supply of consumer goods. In opposition to their predecessors, for whom sociopolitical conflict had been central, historians like David Potter stressed the consensual dynamic of American history; see David Potter, *People of Plenty: Economic Abundance and the American Character* (Chicago: University of Chicago Press, 1954).

The Progressive tradition was revived and transformed in the 1960s and 1970s, however, by the revisionist historians of the New Left, who adopted a less Manichaean approach than their predecessors. See, for example, David W. Noble, "The Reconstruction of Progress: Charles Beard, Richard Hofstadter and Postwar Historical Thought," in Larry May, ed., *Recasting America: Culture and Politics in the Age of the Cold War* (Chicago: University of Chicago Press, 1989), 61–75. The early 1960s marked a rather clear-cut turning point, for it was then that the great paradigms of American achievements began to be questioned. Radical historians indicted industrial society. Star-spangled banners were burning at Berkeley, the civil rights movement hardened its lines, and students demonstrated against the Vietnam War and the imperialism of a country that did not seem to lag behind the colonial powers of the past. Big business stood accused of being part of an oppressive system responsible for the repression of the labor movement, the segregation of black people, and the misery of the Third World—all the doing of cynical multinationals.

In the early 1960s, Rachel Carson denounced the ecological ravages caused by industry in three articles in the *New Yorker*, which were reprinted in her book *Silent Spring* (Boston: Houghton Mifflin, 1962). Carson decried not only the silencing of nature, deprived of its natural sounds and smells, but also the smoke spewed forth by chemical factories and their hideous assemblies of vats, towers, and conduits, which banefully crisscrossed American landscapes. A devastated environment seemed about to become the norm. The impact of *Silent Spring* was profound, and some 600,000 copies were sold in less than a year and a half. On Rachel Carson, who died prematurely in 1964, see Paul Brooks, *The House of Life: Rachel Carson at Work* (Boston: Houghton Mifflin, 1972). Heir to a profoundly American tradition and a great lover of nature—at times she sounds like Thoreau—Carson also tackled other highly contemporary problems, such as the radioactive fallout from nuclear tests. See Samuel P. Hayes, *Beauty, Health and Permanence: Environmental Politics in the United States, 1955–1985* (New York: Cambridge University Press, 1987). In the 1963 CBS Special "The Silent Spring of Rachel Carson," a televised debate between Carson and a representative of Dow Chemical, the issue of pesticides was suddenly projected into the limelight.

Silent Spring was followed by other books that criticized the very principle of the great technological systems, considering them the cause of the country's cultural, social, and environmental iniquities. On the origin of the environmental movement, see

Donald Fleming, "Roots of the New Conservation Movement," *Perspectives in American History* 6 (1972): 7–91.

The chemical industry in particular was targeted for manufacturing napalm for the Vietnam War, pesticides, and other polluting chemicals ravaging America. A discussion of these topics is found in Thomas P. Hughes, *American Genesis: A Century of Invention and Technological Enthusiasm, 1870–1970* (New York: Viking, 1989), chapter 9; see also id., ed., *Changing Attitudes Toward American Technology* (New York: Harper & Row, 1975), a collection of some of the most important articles published in the 1960s. And see, too, Yaron Ezrahi, Everett Mendelsohn, and Howard Segal, *Technology, Pessimism, and Postmodernism* (Dordrecht: Kluwer Academic Publishers, 1993), and Steven L. Goldman, ed., *Science, Technology, and Social Progress* (Bethlehem, Pa.: Lehigh University Press, 1989).

Numerous articles have recalled the dark record of the chemical industry in the twentieth century: the involvement of the German firm IG Farben in the Nazi Holocaust, DuPont's building of the atom bomb, environmental pollution and disasters like that at Bhopal, India, in December 1984, when a Union Carbide plant leaked toxic gas that killed thousands outright and injured hundreds of thousands. All this has given the industry an exceedingly bad press to this day: only 7 percent of Americans judge it favorably, compared to 23 percent for all other industries, despite the well-orchestrated public relation campaigns launched by chemical companies. In the United States, the chemical industry ranks twenty-third out of twenty-four on the scale of popularity for all industries in 1992, only slightly ahead of the nuclear industry. See *News of the Beckman Center for the History of Chemistry* 9, 1 (Spring 1992): 0-2–0-3.

The collapse of the Soviet Union revealed ecological ravages of unique scope, awakening Americans anew to the dangers of chemical and nuclear pollution. From the Urals to Siberia, from the Barents Sea to Lake Baikal, the Russian ecological disaster turned out to be even more outrageous than anyone could have imagined. The radioactive ship cemeteries of Murmansk, the nuclear submarines sunk in the Baltic, the mutant pine trees of Tchernobyl, the massive spreading of dioxins by military plants engaged in the manufacture of chemical weapons, and the proliferation of various cancers and rapid aging syndrome—all of this has made the front pages of the press in recent years. There are pictures of fishermen of Lake Baikal returning to port empty-handed and weeping by the waters on which they had plied their trade for a half-century. In the Promethean and constructivist delirium of communism, which never considered nature anything but material to be manipulated, chemical manufacturing was without a doubt the industry that most surely and most massively polluted and devastated Russia. But the United States was not free from the ravages of industrial chemistry either, for even though here, no Private set out to change the world, chemical and nuclear dumps exist in abundance, and emissions of polluting gases have been widespread and were at times uncontrolled, at least until the creation of the U.S. Environmental Protection Agency in 1970. Exposés have appeared about "Plutonium City," the Hanford site in Washington State, where plutonium was produced during

World War II by DuPont and later by General Electric. See Michelle Gerber, *On the Home Front: The Cold War Legacy of the Hanford Nuclear Site* (Lincoln: University of Nebraska Press, 1992); Michael d'Antonio, *Atomic Harvest: Hanford and the Lethal Toll of America's Nuclear Arsenal* (New York: Crown, 1993). And see also Peter Goin, *Nuclear Landscapes* (Baltimore: Johns Hopkins University Press, 1991), and James M. Fallows, *The Water Lords: Ralph Nader's Study Group Report on Industry and Environmental Crisis in Savannah, Georgia* (New York: Grossman, 1971).

The chemical industry—rather like its younger sister, the nuclear industry—has alternately been the object of lavish praise and deep loathing, both informed by underlying strong emotional and political commitments, from which historians are no more exempt than anyone else. However legitimate ideological extremes may be, one should try to avoid them. It will not do simply to praise the good the chemical industry has done for humanity by writing obliging chronicles featuring aspirin and nylon; or, conversely, to denounce its necessarily nefarious character. I am not about to plead for neutrality, as if that would make my investigation rational. There is no contradiction between being rational and being attuned to the point of view of others, in particular that of the forgotten, the victims, the vanquished. "The opposite of emotional is not 'rational' but the inability to be moved," Hannah Arendt observes in *On Violence* (New York: Harcourt Brace Jovanovitch, 1969), 64. Indignation, however, can be no more than a starting point for mobilization; it does not provide a method of investigation. This being the case, it is useful to turn to the work of business historians, not so as to break away from the demands of emotions and politics, but in order to answer the twin questions "What should be done?" and "How should it be done?"

Chandlerian Perspectives

From the 1960s on, historians of business, foremost among them Alfred D. Chandler Jr., put moral and political appreciations of the activities of major companies to one side in order to study their economic growth (industrial and commercial strategies, internal organization). See Alfred D. Chandler Jr., *Strategy and Structure: Chapters in the History of the American Enterprise* (Cambridge, Mass.: MIT Press, 1962); Thomas C. Cochran, *American Business in the Twentieth Century* (Cambridge, Mass.: Harvard University Press, 1972); and Glenn Porter, *The Rise of Big Business in the Twentieth Century* (New York: Crowell, 1973). Inspired by Joseph Schumpeter's pioneering studies of the dynamic of modern capitalism, encouraged by the entrepreneurs, and stimulated by recent innovations, they analyze in great detail the structure of firms and the manner in which directors and mid-level managers organized commercial and manufacturing activities in keeping with the market. By means of thorough case studies, Chandler highlights the role of managers and engineers in the running of large enterprises in the twentieth century. In following him, business historians gave up the critical gaze of their predecessors—and with it, much that was valuable in the independent positions of both Progressive and revisionist historians.

One of the most important aspects of Chandler's work—the use of a comparative method is one of his trademarks—was his detailed analysis of the differences between the small businesses of the period before the Civil War and the giant new companies of the late nineteenth century that were run by salaried employees. See Alfred D. Chandler Jr., *The Visible Hand: The Managerial Revolution in American Business* (Cambridge, Mass.: Belknap Press of Harvard University Press, 1977), chapter 9. A first point was the enormous capital represented by the new companies. United States Steel, founded in 1901, was capitalized at the level of $1 billion, whereas before 1860, very few companies had had a capital above $1 million. A second point was their organization, for the major modern company brought together in one and the same structure various operational units directed and coordinated by salaried employees, rather than by the "invisible hand of the market." Thirdly, these large companies were set up to produce on a very large scale to amortize the enormous fixed capital invested. The vast majority of these big companies engaged in high-technology activities, and this was no coincidence, for the constitution of large horizontally and vertically integrated groups had to be based on the highest possible threshold of entry into the market in order to discourage competition. New industries were therefore concentrated in high-technology sectors such as electricity, metallurgy, or chemistry, and remained rather sparse in low-technology sectors such as paper, furniture, or printing. See Alfred D. Chandler Jr., with the assistance of Takashi Hikino, *Scale and Scope: The Dynamics of Industrial Capitalism* (Cambridge, Mass.: Belknap Press of Harvard University Press, 1990). For this reason, technical innovation and the development of the most efficient manufacturing procedures possible were of central importance to the leaders of large companies, for they were necessary to the very survival of the company. "Modern business enterprise was thus the institutional response to the rapid pace of technological innovation and increasing consumer demand in the United States during the second half of the nineteenth century" (Chandler, *Visible Hand*, 12).

These concerns were particularly vital in sectors whose very existence depended on the application of new scientific knowledge. Above all, this was the case in the electrical and communications industries, and to a lesser extent in the chemical industry as well. It is therefore understandable that historians of business, and in their wake those of technology, took an early interest in the first steps of the electrical industry and the telegraph and telephone industry. See Leonard Reich, *The Making of American Industrial Research: Science and Business at GE and Bell, 1876–1926* (New York: Cambridge University Press, 1985); W. Bernard Carlson, *Innovation as a Social Process: Elihu Thomson and the Rise of General Electric, 1870–1900* (New York: Cambridge University Press, 1991); Thomas P. Hughes, *Networks of Power: Electrification in Western Society, 1880–1930* (Baltimore: Johns Hopkins University Press, 1983); and George Wise, *Willis R. Whitney, General Electric and the Origins of U.S Industrial Research* (New York: Columbia University Press, 1985). These new activities marked the opening of a new industrial era, characterized by technical practices founded on scientific knowledge derived from the physics and chemistry of the atom, the molecules, light, and magnetism. Even Edison, that legendary prototype of the maverick inventor, working at night in his laboratory den in Menlo Park, New Jersey, was in reality surrounded by high-powered physicists and chemists. See Hughes, *American Genesis.* "Invention then becomes a branch of business, and the application of science to immediate production [aims] at determining the inventions at the same time it solicits them," Karl Marx observes in *The Grundrisse,* ed. and trans. David McLellan (New York: Harper & Row, 1971), 140.

It was at Chandler's suggestion that historians of technology, foremost among them Thomas Hughes, began in the early 1960s to look into the relationship between intensified techniques and industrial growth. Accordingly, these historians studied the creation of the first American industrial research laboratories—at General Electric, created in 1900, and at Bell and DuPont, dating from 1902—that brought together high-powered physicists and chemists as in the laboratories of the great German chemical firms BASF, Bayer, and Hoechst. They also examined, albeit not quite as thoroughly, the appearance of the new electrical and chemical engineers.

Although they replaced Chandler's analysis of the interactions between a company's strategies and its internal organization with that of the relation between industrial strategies and scientific and technological investments, these historians of innovation by and large remained within Chandler's framework. Leonard Reich, for instance, distinguished between two strategic approaches used by AT&T: the proactive, which consisted of developing new products and procedures; and the defensive, which aimed to protect its innovations by erecting a "wall" of patents to block any competitor. See Reich, *Making of American Industrial Research.*

This approach fitted perfectly into Chandler's model, since Reich was interested in showing how the communications company raised the threshold of entry into the market so high that it was out of reach for most of the competition, and hence how AT&T maintained its "oligopolistic" position in the long-distance communications market. Having rebelled—and rightly so—against the overly externalist perspectives of their predecessors of the Progressive school, historians of business and industrial research failed to go beyond the strictly internal perspectives of the companies under consideration. This made for frequently repetitive research topics, which in the end amounted to little more than examining the relationship between growth and innovation.

To take just one example, the reader of *Science and Corporate Strategy: Du Pont R&D, 1902–1980,* by David A. Hounshell and John Kenly Smith Jr. (New York: Cambridge University Press, 1988), which was commissioned and financed by DuPont, practically never gets out of the company's laboratories. These authors intend to show how research has become an important part of the chemical firm's general strategy, and how its laboratories were reorganized on a regular basis in response to the imperatives of the moment. Exclusively focused on the firm, this approach does not offer a new way of doing history but, rather, deals with the kind of topic pursued by management science. Too many historians of industrial research have allowed management specialists and the protagonists of the story to define their topics. This historiography is dominated by questions about the best way to manage research in very capital-intensive and advanced-technology fields. Yet the historical discourse must aspire to be something more than management analysis—even at the high level to which Chandler has raised it. This is not meant to be a criticism of management science, to which historians should probably pay more attention. What I wish to stress is that the historian's discourse belongs to a different category than that of the management expert.

One possible way to do this is to go beyond monographic work and the study of organization. We must take into account interactions among organizations as well as between organizations, society, and politics. Such interactions blur the boundaries between what is "internal" and "external" to a company. Insofar as laboratories, factories, and offices are dependent on great technical systems, the interlocking of different institutions, and also the increased weight of the public powers in the country's social and economic life, historians should focus on the porosity of organizations. To Chandler, the government is at most only one variable. No wonder he himself has admitted inability to fully comprehend "the shape of post-modern, super-modern super-enterprises that may replace the industrial enterprise." See Alfred D. Chandler Jr., "Response to the Contributors to the Review Colloquium on *Scale and Scope*," *Business History Review* 64, 4 (1980): 744.

The history of DuPont has been intensively explored by Chandler, and subsequently by Hounshell and Smith. As a family member himself, Chandler took an early interest in DuPont. Following his *Strategy and Structure* (1962), which includes a chapter on DuPont, he and Stephen Salsbury co-authored a biography of its legendary head, *Pierre S. du Pont and the Making of the Modern Corporation* (New York: Harper & Row, 1971). Chandler subsequently also devoted substantial chapters to DuPont in other books, analyzing its successive reorganizations in 1910 and 1920 and showing that it was in the vanguard of modern management techniques in America. Chapters devoted to DuPont are found in all of Chandler's books, from *Strategy and Structure* to *Scale and Scope*. Hounshell and Smith, for their part, studied the history of the firm's research laboratories and showed how research was constantly reorganized in keeping with the company's overall strategy at the instigation of individual directors. See Hounshell and Smith, *Science and Corporate Strategy*, particularly chapters 1, 5, 12, 16, and 25. Although their approach differs from Chandler's, they address similar issues. The "political" history of DuPont's relations with other institutions has yet to be written.

The history of the chemical industry suffers from some of the defects of Chandler's approach, including a disregard for historical concepts, as John K. Smith points out in "The Evolution of the Chemical Industry: A Technological Perspective," in Seymour Mauskopf, ed., *Chemical Sciences in the Modern World* (Philadelphia: University of Pennsylvania Press, 1993), 137ff. Most of the existing studies are focused on particular firms or products, without a larger perspective. Written by chemists or engineers, they are technically proficient but insufficiently historical. The histories of chemical companies are often official histories, encouraged or even commissioned by the firms studied, and as such they respond only to issues raised within these companies—or, worse yet, are integrated into their public relations strategies. A characteristic example is that of Standard Oil of New Jersey, which asked several historians to write a history of the firm, which was subsequently published in three volumes: Ralph W. Idy and Muriel E. Idy, *Pioneering in Big Business, 1882–1911: History of the Standard Oil Company* (New York: Harper & Row, 1955); George S. Gibb and Evelyn H. Knowlton, *The Resurgent Years,*

1911–1927: History of the Standard Oil Company (New York: Harper & Row, 1956); and
H. M. Larson, E. H. Knowlton, and C. S. Popple, *New Horizons, 1927–1950* (New York:
Harper & Row, 1971). Even Hounshell and Smith's *Science and Corporate Strategy*—remarkable as it is in many respects—is interested above all in addressing the questions
asked by DuPont's strategists in the mid 1980s, a time when the firm's research program
was showing serious signs of running out of steam, and when a historical perspective
might well have been a valuable aid in redefining the major goals of research and development for the rest of the century.

The few academic studies in the history of the chemical industry are usually structured around the two Chandlerian concepts of scale and scope. Scale has to do with
economies of scale, which are crucial in certain sectors (especially inorganic chemistry), but constitute a problem for the engineer, since the temperatures and pressures
used in laboratory experiments cannot easily be transferred to large-scale operations.
Scope refers to the diversifications arising from an initial product. The more chemicals
a company produces, the more new ones it can conceive. The combination of economies of scale and diversification, with their consequences in terms of industrial production and commercial structures, is supposed to determine the general evolution of
the industry. John K. Smith suggests hitherto unexplored areas of research, such as the
design of chemical plants, catalytic chemistry, and the history of equipment (Smith,
"Evolution of the Chemical Industry," 139). He does not, however, propose any opening
that could bring this research area out of its isolation. It has become urgent to break
free from an exclusive concern with the three related variables of market, innovation,
and growth.

Recreating a Context

It would be useful to give the history of industrial innovation the conceptual autonomy
it lacks in the Chandlerian version of business history. Several attempts in this direction
have been made in recent years. A first possibility consists of looking at technology as
a social construction, an approach that has been known for twenty-five years under
the acronym SCOT (social construction of technology). Growing out of social studies
of knowledge, this approach has powerfully transformed the history of techniques by
endeavoring to show in what ways technical objects and practices reflect and act upon
social, political, and cultural conditions. See Wiebe E. Bijker, ed., *The Social Construction of Technological Systems: New Directions in the Sociology and History of Technology*
(Cambridge, Mass.: MIT Press, 1987). The main theoretical interest of SCOT is that
it endows knowledge, production, consumption, even objects, with social meanings.
Often the result of microhistorical examinations, SCOT has brought about a profound
renewal in a history of technology that had heretofore been caught up in determin-
ist and positivist thinking. Business historians can certainly learn a great deal from
recent histories of technology, such as Thomas Hughes's work on the great technological systems, especially his books *American Genesis* and *Networks of Power*, where
Hughes keeps track of all the multiple aspects of innovation. Also important are the

studies of the interchangeability of parts and on the assembly line, both of which have enriched our understanding of the "rationalization" of production in the first decades of the twentieth century. Among many references, particular attention should be paid to David A. Hounshell, *From the American System to Mass Production, 1800–1932: The Development of Manufacturing Technology in the United States* (Baltimore: Johns Hopkins University Press, 1984); Yves Cohen, "Inventivité organisationelle et compétitivité: L'interchangeabilitité des pièces face à la crise de la machine-outil en France autour de 1900," *Entreprises et Histoire* 5 (June 1994): 53–72; Aimée Moutet, *Les logiques de l'entreprise: La rationalisation dans l'industrie française de l'entre-deux-guerres* (Paris: École des hautes études en sciences sociales, 1997); and Patrick Fridenson, "Un tournant taylorien dans la société française (1904–1918)," *Annales. Histoire, Sciences Sociales* 42, 5 (September–October 1987): 1031–60.

A second possibility consists of thinking in terms of a cultural, even anthropological, history of technical innovation. Jeffrey L. Meikle engages in such a study in *American Plastic: Molding a Culture of New Materials* (New Brunswick, N.J.: Rutgers University Press, 1995), which looks at the consumers of mass-produced chemical products (articles made of Bakelite, nylon, and various plastics) and at those consumers' possible influence on production in view of a cultural history of modernity. Susan Douglas, *Inventing American Broadcasting, 1899–1922,* (Baltimore: Johns Hopkins University Press, 1987), looks at radio in the United States and the cultural impact of the programs broadcast. The major interest of these studies is to show that innovation, far from being confined to the sphere of production, is also initiated by advertisers, distributors, and consumers themselves.

Another way of going beyond scale and scope is to come to grips with political and social factors outside the sphere of the company as such. In doing so, one enters the realm of political and social business history. Without returning to Progressive history and certain of its moral judgments, it does seem fruitful to reinsert these great organizations into a large historical context and to make use of social-science concepts and methods. In *Making America Corporate, 1870–1920* (Chicago: University of Chicago Press, 1990), Olivier Zunz has shown, for instance, how a new class of employees, engineers, and managers became invested in the development of major companies and the construction of a continentwide economy, and how managerial capitalism was able to create a new culture of work that gave rise to the famous twentieth-century American middle class. Another historical approach consists of concentrating on one socioprofessional group at work in the organization.

Generally speaking, this kind of study is part of the history and sociology of the professions. Initially, the sociology of the professions was greatly influenced by Talcott Parsons, who in his *Essays in Sociological Theory* (New York: Free Press, 1954) attributed professional status to the four groups of engineers, physicians, professors, and lawyers. According to Parsons, the professional differentiates himself from both the scientist and the technician by simultaneously putting his knowledge to work and keeping abreast of new developments in his field. This functional and prescriptive view presented an ideal type of professional, who, motivated by ethical choices and functioning as the agent of an institutionalization of the social functions, was not affected by historical change. Hence the interest of Everett Hughes's *The Sociological Eye* (1971; repr., New Brunswick, N.J.: Transaction Books, 1984), which concentrates on the social standing of the professions and on the manner in which they negotiate their standing within the different social institutions. Hughes has shown that professionals have long sought to define the extent of their competencies and the value society placed on their services. Having become more interested in interaction than in norms, the sociology of the professions now began to consider the professions as social entities in permanent evolution within ever-changing boundaries. Most of the sociologists of the professions have followed in Hughes's footsteps, although Eliot Freidson, in *Professional Powers: A Study of the Institutionalization of Formal Knowledge* (Chicago: University of Chicago Press, 1986), also stresses the role of the professional associations, and Andrew Abbott, in *The System of Professions* (Chicago: University of Chicago Press, 1988) stresses the importance of controversies over the legal boundaries of professional groups. See also the excellent overview by Eli Commins, "Pratiques sociales et culturelles des médecins et avocats de New York, 1900–1960" (Diplôme d'études approfondies thesis, École des hautes études en sciences sociales, 1999). In fact, these questions of boundaries have been of central concern in the engineering profession.

The history of engineering in the United States is roughly divided into two major categories. The first is the study of engineering associations: the American Society of Civil Engineers, the American Society of Mechanical Engineers, the American Institute of Chemical Engineers, and the American Institute of Electrical Engineers. On these, see Terry S. Reynolds, *75 Years of Progress: A History of the American Institute of Chemical Engineers* (New York: AIchE, 1983); id., ed., *The Engineer in America: A Historical Anthology from Technology and Culture* (Chicago: University of Chicago Press, 1991); Monte A. Calvert, *The Mechanical Engineer in America, 1830–1910: Professional Cultures in Conflict* (Baltimore: Johns Hopkins University Press, 1967); Bruce Sinclair, *A Centennial History of the American Society of Mechanical Engineers, 1880–1980* (Toronto: University of Toronto Press, 1980); William H. Wisely, *The American Civil Engineer, 1852–1974: The History, Traditions, and Development of the American Society of Civil Engineers Founded in 1852* (New York: ASCE, 1974); and A. Michael McMahon, *The Making of a Profession: A Century of Electrical Engineering in America* (New York: IEEE, 1984). These studies are based on the archives of these associations and were officially commissioned by them. Although rich in information, they do not always take a sufficient distance from their subject (significantly, Reynolds's book about the American Institute of Chemical Engineers is titled *75 Years of Progress*). More important, such studies take a purely institutional and statutory approach to the engineering profession. When it comes to the older specialties (civil and mechanical engineering), they essentially deal with the manner in which they were structured in the late nineteenth century, when the growing power of big companies threatened the traditional forms of this occupation, and with the manner in which the Progressive movement spread in these associations, especially that of the mechanical engineers, where the disciples of Frederick Taylor were active. For the newer specialties (electrical and chemical engineering), these studies show that the creation of analogous associations responded to the need for institutional and professional recognition.

The second category of works on the history of American engineering consists of studies grounded in the history of technology. These focus on the engineers' contribution to a specific company or a specific project, or else emphasize the creative genius of one of the great engineers. Ronald R. Kline's *Steinmetz: Engineer and Socialist* (Baltimore: Johns Hopkins University Press, 1992), a biography General Electric's chief engineer, Charles Steinmetz, followed on the heels of two other notable biographies, Thomas P. Hughes's *Elmer Sperry: Inventor and Engineer* (Baltimore: Johns Hopkins University Press, 1971) and Stuart Leslie's *Boss Kettering: Wizard of General Motors* (New York: Columbia University Press, 1983). Let us pause for a moment on Hughes's biography of Sperry. Although this work was suggested by Chandler, Hughes focuses not so much on Sperry's entrepreneurial dimension as on his qualities as a great creative engineer. As Hounshell has rightly pointed out, this biography is quite different from Chandler and Stephen Salsbury's *Pierre S. du Pont and the Making of the Modern Corporation,* whose actual subject is not so much the head of the company as the company itself. The genesis of these two works is narrated by David A. Hounshell in "Hughesian History of Technology and Chandlerian Business History: Parallels, Departures, and Critics," *History of Technology* 12 (1995): 205–24. In *Elmer Sperry,* his first major book, Hughes was above all interested in Sperry's brilliant technical inventiveness (except in one chapter, "Brain Mill for the Military," which was fittingly included in *American Genesis* fifteen years later).

To my mind, only one book in the history of engineering does not fit into either of these historiographical categories. This is David F. Noble's *America by Design: Science, Technology, and the Rise of Corporate Capitalism* (New York: Knopf, 1977), which sets out to show how the engineers have been the agents of big capital and how their work has been subjected to its imperatives. In the same Marxist vein, Noble's *Forces of Production: A Social History of Industrial Automation* (New York: Oxford University Press, 1984) shows that the numerically controlled machine tools that came into use in the early 1950s were in no way better than competing systems, but that they won out because they transferred control of the work on the shop floor to the office of the engineer and his military customers and thus reduced their dependence on skilled workers, who were often unionized and always ready to strike.

Rather than looking at a profession as a whole or one particular figure in the world of American engineering, the present book seeks to draw a group portrait, not so much from a social and cultural as from a political perspective. The other major possibility of breaking away from the Chandlerian approach is to explore the relationship between technology and politics, that is to say, to examine the ways in which the organization of the public powers, their political choices, and their relations with the world of private production have both shaped and reflected technology and its originators. In this respect, important work has been done by historians of technology specialized in the study of the Cold War, a particularly rich area for the study of military technology. Thus the sociologist Donald McKenzie, in studying intercontinental guided missiles, has underscored the rival claims by the different branches of the military over ground control stations. For her part, Pamela E. Mack, in *Viewing the Earth: The Social Construction of the Landsat Satellite System* (Cambridge, Mass.: MIT Press, 1990), has analyzed the manner in which NASA's Landsat satellite program has from its very conception incorporated the contradictory specifications of the different interest groups that sponsored it. Stuart Leslie's *The Cold War and American Science: The Military Industrial Academic Complex at MIT and Stanford* (New York: Columbia University Press, 1993) similarly shows to what extent the teaching programs of major universities such as MIT or Stanford—and hence the scientific fields themselves—were defined in response to the needs of the U.S. Army between 1940 and the 1970s. Yet business history has not become involved in this field of research focused on the universities and the military in the context of the Cold War and massive military spending.

The most politically aware historians of organizational synthesis, foremost among them Louis Galambos, have given us analyses that do integrate the political dimension of production organizations. In a synthetic work, *The Rise of the Corporate Commonwealth: United States Business and Public Policy in the 20th Century* (New York: Basic Books, 1988), Galambos and Joseph Pratt stress the regulatory function of the federal government and the different agencies that were created for this purpose, as well as the gradual acceptance of the economic and social functions of the government on the part of big business. Yet here, too, the approach remains macroeconomic and macropolitical. Robert Collins, in *The Business Response to Keynes, 1929–1964* (New York: Columbia University Press, 1981), which studies the National Association of Manufacturers, has emphasized that after World War II, this conservative employers' organization converted to a moderate form of Keynesianism and to government intervention in order to prevent cyclical downturns of the economy.

These historians have for the most part remained close to a very institutional level of analysis that does not venture out into technical or social details. Aspects related to technology, in particular, are relegated to the background. But I believe that they are determining factors, in the sense that the history of technology advanced in step not only with high capitalism, as Noble would put it, but also with what Galambos calls the "corporate commonwealth," that is, the system of public and private partnerships that has characterized America in the twentieth century. This system was gradually built up from the early years of the century, and this development was accelerated by the two world wars, especially the second. At the end of the 1940s, in the context of the Cold War, this system experienced its heyday as the incarnation of what I call "the ideology of prosperity and security."

Analyzing a corporation in its relations with the "outside," that is, with knowledge- and production-based organizations, as well as with political institutions, thus appears to be a fruitful approach. In such an undertaking, the history of industrial innovation is assigned a new role. Instead of being strictly related to a particular company's commercial and industrial strategy, it is here considered from a more political angle. Technology is not the inevitable effect of the laws of nature and of a technical necessity that imposes its specific forms. Quite the contrary, it is dependent on a given political and social order. That is why we must not establish too rigid a separation between technology, on the one hand, and society, politics, and culture, on the other. Technology is not a collection of technical objects and know-how; it exists as a network of social connections, so that studying it enables us to deal with issues of concern to students of social, cultural, and political history. All in all, then, it seems desirable to rethink the history of industry, and particularly that of the chemical industry, as integral parts of general history, and to go beyond issues of management.

In attempting this kind of "open" history, it was necessary to choose one professional group, one industry, and one company that would allow me to highlight this relational dimension of twentieth-century American organizations and the manner in which the multipolar network of public and private initiatives slowly coalesced. I chose the chemical engineers, the chemical industry, and DuPont.

Index

Page numbers followed by an *f* or *t* indicate figures and tables.